Karl Ernst von Baer [1792-1876] Anton Dohrn [1840-1909]

Correspondence

TRANSACTIONS

of the

American Philosophical Society

Held at Philadelphia for Promoting Useful Knowledge

VOLUME 83, Part 3, 1993

Karl Ernst von Baer [1792–1876] Anton Dohrn [1840–1909]

Correspondence

edited by CHRISTIANE GROEBEN
Stazione Zoologica 'Anton Dohrn,' Naples

With an introduction by JANE M. OPPENHEIMER
Professor Emeritus of Biology and the History of Science
Bryn Mawr College

Translations from the German
by CHRISTIANE GROEBEN *and* JANE M. OPPENHEIMER

THE AMERICAN PHILOSOPHICAL SOCIETY

Independence Square, Philadelphia

1993

Library of Congress Catalog
Card Number: 92-75293
International Standard Book Number 0-87169-833-1
US ISSN 0065-9746

TABLE OF CONTENTS

Introduction 1

Editorial Remarks and Acknowledgments 25

List of Abbreviations 28

Chronological List of the Letters 29

Letters 1–41 31

APPENDIX: Briefe 1–41 97

Bibliography 151

Index of Names 155

ILLUSTRATIONS

Figure 1 Anton Dohrn, photograph (1863) .... *page* 36

2 The Zoological Station in 1873, photograph 49

3 Anton Dohrn, study in oil (1873) 53

4 Karl Ernst von Baer, bust (1873) 57

5 Karl Ernst von Baer, photograph (1859) 67

6 Karl Ernst von Baer, photograph (1872?) 68

7 Karl Ernst von Baer, photograph (1874) 72

INTRODUCTION

In 1869 Anton Dohrn wrote to Karl Ernst von Baer to enlist his help in supporting the Stazione Zoologica that would soon be set up in Naples (letter #4 in this collection). This was the first of an exchange of 36 letters between Anton Dohrn and von Baer that constitutes the principal portion of this correspondence. Twelve letters from von Baer to Dohrn have for some time been known to exist in the Dohrn-Archiv in Naples; very recently 21 letters from Dohrn to von Baer have surfaced in the University Library in Giessen (a few others from the same collection, related to the main correspondence, are also presented here). Thus we have an almost complete record of this correspondence, wherein one extant letter replies to another.

Von Baer's first biographer was Ludwig Stieda, an anthropologist, a professor of anatomy in Dorpat in 1876 when von Baer died there in retirement.[1] Stieda had been entrusted by von Baer's family with the latter's scientific correspondence. When Stieda left Dorpat he took it with him first to Königsberg, and then to Giessen where he retired. From then on the whereabouts of this correspondence was long unknown. Shortly before his death, Alexander Stieda, one of Ludwig's sons, made known to Heinrich von Knorre, the translator into German of Raikov's Russian biography of von Baer, that during the last month of World War I in 1918 Ludwig Stieda's whole library, including von Baer's scientific correspondence, was destroyed by enemy fire. Much to the surprise of many, during the second half of 1971 manuscripts constituting the scientific correspondence of von Baer came to light in the University of Giessen Library.[2] In it were included the letters to him from Carl August, Heinrich, and Anton Dohrn that we include here.

---

1 The principal published sources of biographical information about von Baer are the following: L. Stieda, *Karl Ernst von Baer. Eine biographische Skizze*, 2d ed. (Braunschweig: F. Vieweg, 1886; unchanged from 1st ed., 1877); B. E. Raikov, *Karl Ernst von Baer 1792–1876. Sein Leben und sein Werk*, trans. H. von Knorre (Leipzig: Johann Ambrosius Barth, 1968); K. E. von Baer, *Nachrichten über Leben und Schriften des Herrn Dr. Karl Ernst von Baer, mitgetheilt von ihm selbst* (St. Petersburg: Ritterschaft Ehstlands, 1865, distributed privately; trade editions published under same title: St. Petersburg: H. Schmitzdorff, 1866, and Braunschweig: F. Vieweg, 1886). English translation of 1886 edition: K. E. von Baer, *Autobiography of Dr. Karl Ernst von Baer*, ed. Jane M. Oppenheimer, trans. H. Schneider (Canton, Mass.: Science History Publications, 1986); *Folia Baeriana*, ed. T. Sutt, 4 vols. (Tallinn: Valgus, 1975, 1976, 1978, 1983); vol. 5, ed. T. Sutt and V. Kaavere (Tallinn: Valgus, 1990). *Perepiska Karla Bera*, ed. T. A. Lukina, 4 vols. (Leningrad: Nauka, Leningrad Section, 1970, 1975, 1976, 1978); H. von Knorre and H. Schierhorn, 1975, "Karl Ernst von Baer (1792–1877). Eine ikonographische Studie," in *Beiträge zur Geschichte der Naturwissenschaften und der Medizin, Festschrift für Georg Uschmann, Acta Historica Leopoldina* 9 (1975): 227–268.

2 H. von Knorre, "Die Entstehungsgeschichte von K. E. Baers 'Sendschreiben': De ovo

When Theodor Heuss created his definitive and highly detailed biography of Anton Dohrn,[3] he mentioned Karl Ernst von Baer only cursorily. He spoke of him as an embryologist of distinction; he said without elaboration that von Baer helped to secure Russian Imperial support for the marine station that Dohrn was developing, and finally, toward the end of the biography, in a discussion of Dohrn's then recently developed theories relating to the evolution of vertebrates from invertebrates, he devoted nearly a whole page to some of von Baer's beliefs and non-beliefs about Darwin's theories. Here Heuss permitted von Baer to speak for himself by quoting some passages from a letter that von Baer wrote to Dohrn in June 1875 at the age of 83.[4]

The correspondence that we present here sheds a bright light on a number of important aspects of the actions and thoughts of two scientists whose contributions were significant to the development of biology in their own times. We continue to benefit from them today more than a century later, but that is irrelevant to our purpose here. Their exchange of ideas as expressed in these letters has an immediacy that no biography by others can bring to our perception of their qualities. They were privileged to enjoy, alas only through correspondence, an intimacy that shines through their words and brightly illuminates the personalities of both.

As we said at the start, Anton Dohrn first wrote to von Baer in 1869. The previous year, 1868, had been a decisive one in Dohrn's life; he was then 28 years old. His first interest as a maturing youth had been in entomology; he had then thought he might become a zoologist, but subsequently his plans for his career vacillated. The subject of zoology became at one point uninteresting to him, but in 1862, he became acquainted with Darwin's ideas on the origin of species. This was the

---

mammalium et hominis genesi 1827 und vier Briefe Karl Ernst von Baer [*sic*] an Carl Asmund Rudolphi," *Mitt. d. deutsch. Akad. d. Naturforscher Leopoldina* 17 (1971): 237–286; see "Nachwort" (1972), 266–267, regarding the fate of von Baer's scientific correspondence.

[3] The definitive biography of Dohrn is T. Heuss, *Anton Dohrn* (Berlin-Zürich: Atlantis, 1940; 2nd enl. ed. Stuttgart-Tübingen: Rainer Wunderlich, 1948; English translation, Heidelberg, Berlin: Springer-Verlag, 1991). The second edition includes an essay by Margret Boveri, "Das Haus am Rione Amadeo," describing life in the Dohrn household. *Naturwissenschaften* 28, no. 51, 1940, commemorated the hundredth anniversary of Dohrn's birth: it includes, among a number of other articles, a reprint of the perceptive memorial address on Dohrn delivered by Theodor Boveri at the International Zoological Congress in Graz, previously published separately by Boveri as *Anton Dohrn. Gedächtnisrede, gehalten auf dem Internationalen Zoologen-Kongress in Graz am 18. August 1910* (Leipzig: S. Hirzel, 1910). An excellent brief evaluation of Dohrn is that by E. B. Wilson, "The Memorial to Anton Dohrn," *Science* 34 (1911): 632–633.

An extensive bibliography of Anton Dohrn's publications (80 items) is found in A. Kühn, "Anton Dohrn und die Zoologie seiner Zeit," *Pubbl. d. Staz. zool. Napoli*, suppl. 1950. K. J. Partsch, *Die Zoologische Station in Neapel. Modell internationaler Wissenschaftzusammenarbeit* (Göttingen: Vandenhoeck & Ruprecht; Studien zur Naturwissenschaft, Technik und Wirtschaft im Neunzehnten Jahrhundert 11, 1980) provides a number of important technical and business details relating to the establishment of the Stazione.

[4] Heuss, *Anton Dohrn*, 223.

decisive event that turned him back to zoology, and in 1868 he became habilitated in zoology at Jena.

Also, in autumn 1868 he visited Messina where among other pleasures, he met the Russian Nicolai Miclucho-Maclay, later to be a good friend (see note 16 in the Letters in English, p. 35 below). This was not his first trip to the seashore; he had traveled to Helgoland three years before, and in 1867 and 1868 he had worked in Millport with David Robertson, an amateur zoologist and founder in 1885 of the Millport Biological Station. But the visit to Messina was different; this time Dohrn took with him a portable aquarium that he had designed with Robertson to permit the circulation of water within it. As a result of his experiences in Messina, he first began to think about establishing a chain of marine stations for biological research. And in Messina he met Marie von Baranowska, who would in due time become his bride.

In what phase of his life was Karl Ernst von Baer in 1868? His deep concentration on embryology was about 40 years behind him, though his ideas about it were still dominant, as they remain today. Born in Estonia, at that time a Russian province, he was a member of the Russian aristocracy. In 1868 he was living quietly in Estonia, in Dorpat (now Tartu) whither he had moved the previous year upon his retirement from the Imperial Academy of Sciences in St. Petersburg; Dorpat had been his university when he earned his medical degree in 1814.

The earliest item to appear in the Dohrn–von Baer exchange of letters was dated 1869; why then have we emphasized the year 1868? This was, as we have said, the year that Anton Dohrn first began to think of establishing marine laboratories. The main subject of the correspondence to follow deals with the establishment of the Stazione Zoologica in Naples, and especially with the important role played by von Baer in gaining support for the venture. Thanks in part to him, the Russian ruling regime was one of the first governments to provide funds in support of the Station.

Earlier Backgrounds for Anton Dohrn's Interest in Marine Biology

When Dohrn transported a portable aquarium to Messina in 1868, public aquaria, at least inland, were beginning to develop their own popularity; private ones had become fashionable in the 18th century.[5] The first public one seems to have been set up in Regent's Park in London (if London is to be considered inland) in 1853; as Dohrn knew, one was established in Hamburg in 1867. [Dohrn had worked at the aquarium in Hamburg in spring 1867.] Brehm was to set one up in Berlin in 1869.

---

[5] The passage here beginning with the phrase *aquaria, at least inland* and terminating below on p. 6 with the words *technical organization of the laboratories* is taken from the following: J. M. Oppenheimer. "Some Historical Backgrounds for the Establishment of the Stazione Zoologica at Naples," in M. Sears and D. Merriman, eds., *Oceanography: the Past* (New York, Heidelberg, Berlin-Springer, 1980), 179–181, with the kind permission of Dr. Sears and Springer-Verlag, Heidelberg.

Thus Dohrn's use of aquaria in Messina was not a wholly isolated phenomenon. . . .

This was hardly the first zoological field trip ever made to the seashore, not even for Dohrn. He himself had gone to Helgoland with Haeckel and others in 1865. . . . Spallanzani had studied the fauna of the Sicilian Coast in 1788; Krohn had studied Sicilian coastal fauna in 1846. . . . Louis Agassiz collected specimens from shallow waters bordering Massachusetts from the U.S. Coast Survey vessel *Bibb* as early as 1847.[6] Perhaps more generally influential on the theoretical thinking of laboratory zoologists of the mid-19th century was the fact that Johannes Müller, as I. Müller tells us, often went down to the seashore with his students.[7] In 1841, with Retzius, he studied *Amphioxus* in Bohusland, Sweden, and on the Kelsen Islands near Göteborg. Beginning in 1845 he traveled to the seashore regularly during his vacations: to the North Sea, the Baltic, the Adriatic, the Ligurian. He developed a strong interest in the study of microscopic marine forms and observed for the first time, among many other important discoveries, the metamorphosis of the pluteus larva into the sea urchin; he collected specimens by using tow-nets.[8] In Messina it had been the opportunity to observe unexpected metamorphoses of crustaceans that was of particular appeal to Dohrn, since by his day, in the Darwinian era, unlike in Müller's, this was viewed as an opportunity to ascertain hitherto unsuspected evolutionary relationships by examining embryos as ancestors.

To vacation at the seashore with nets and even microscopes, collecting in shallow waters from rowboats and even from coast survey vessels, is not to establish a chain of permanent sea-side laboratories, which was Dohrn's first hope. But it was a first step towards setting up a single one. Even here, with respect to the founding of a station, or stations, Dohrn was not completely original, nor did he claim to be. The Belgian zoologist Pierre Jean van Beneden [*sic*; his middle name was Joseph, not Jean. Ed.], father of the now better-remembered cytologist Edouard van Beneden, set up a short-lived station at Ostend in 1842.[9] Van Beneden knew that in England the coastal fauna was being exhaustively studied from a taxonomic point of view; he wished to do the same for that of Belgium. In 1841 he went to Ostend on a fishing vacation; that same year he published his first paper on a marine form, *Hydractinia*.[10] After he set up his small laboratory a year or so later, it was used by a number of others, including Johannes Müller and by de Quatrefages, and P. J. H. de Lacaze-Duthiers.

---

[6] E. Lurie, *Louis Agassiz. A Life in Science* (Chicago: University of Chicago Press, 1960).

[7] I. Müller, "Die Wandlung embryologischer Forschung von der deskriptiven zur experimentellen Phase unter dem Einfluss der Zoologischen Station in Neapel," *Med. Hist. J.* 10 (1975): 191–218.

[8] J. Steudel, "Johannes Peter Müller," in C. C. Gillispie, ed., *Dict. Sci. Biogr.* (New York: Scribner) 9 (1974): 567–574.

[9] A. Kemna, *P. J. van Beneden. La vie et l'oeuvre d'un zoologiste.* (Anvers: Buschmann, 1897).

[10] P. J. van Beneden, "Recherches sur la structure de l'oeuf dans un nouveau genre de Polype (genre Hydractinie)," *Bull. Acad. Roy. Scis. Bruxelles* 8 (Pt. 1) (1841): 89–93.

In France an important attempt to study life along the seashore, partly from the point of view of physiology, had been made even earlier,[11] and V. Audouin and H. Milne-Edwards published as early as 1832 an important work on the littoral life along French shores.[12] Audouin died early and then A. de Quatrefages, who had been at Ostend, became associated with Milne-Edwards in the work; their journeys were written up by the former.[13]

Milne-Edwards next worked with Carl Vogt in Paris, and in 1844–1847 they formulated a plan for investigating a coral island and setting up a station on it that would function for at least several years. According to a letter that Vogt wrote to Dohrn, published in 1872, the plan fell through because "the commander of a man-of-war of the Royal Navy would not submit to the direction of a naturalist."[14] Vogt lived in Nice from 1850 to 1852 and tried then to set up a station at Villefranche, vainly, though all he asked for were some rooms in an empty building. In 1863 when his friend Mateucci [*sic;* correct spelling Matteucci. Ed.], a physicist, became Minister of Public Instruction in Italy, Vogt worked out a plan for erecting a station at Naples that was to include instruction among its efforts. But Mateucci [*sic*] left the government and that was the end of that. Vogt tried for the last time, and failed, at Trieste in 1872. He and Dohrn were close to one another, and Dohrn sought and accepted much advice from him. De Lacaze-Duthiers, who had previously worked at Ostend, did succeed in setting up a station at Roscoff in 1871[15] and later one at Banyuls.

Other movements were stirring too, as far away as the United States of America, as the Stazione opened. The first successful marine *teaching* establishment, the Anderson School of Natural History, was opened by Louis Agassiz on Penikese Island, in Buzzards Bay, in 1873; it survived only for two years.[16] This was not a research venture. As early as 1871 investigators from the newly founded U.S. Fish Commission, established through the efforts of Spencer Baird, worked out of Woods Hole in small boats. This was before the Commission's laboratory there was erected in 1885. Baird had had in mind the possibility of its becoming a sort of summer university, with a group of universities using the facilities for teaching as well as research, but these plans had to be withdrawn for political reasons.[17] The Marine Biological Laboratory at Woods Hole was

---

[11] E. S. Russell, *Form and Function. A Contribution to the History of Animal Morphology* (New York: E. P. Dutton, 1916).

[12] A. Audouin and H. Milne-Edwards, *Recherches pour servir à l'histoire naturelle du littoral de la France; ou Recueil de mémoires sur l'anatomie, la physiologie, la classification et les moeurs des animaux de nos côtes* (Paris: Clochard, 1832).

[13] A. de Quatrefages, *Souvenirs d'un naturaliste,* 2 vols. (Paris: Charpentier, 1864).

[14] A. Dohrn, "The Foundation of Zoological Stations," *Nature* 5 (1872): 277–280; quotation from p. 277.

[15] S. Schlee, *The Edge of an Unfamiliar World. A History of Oceanography* (New York, E. P. Dutton, 1973). The correct date for the opening of Roscoff is 1872 [Ed.].

[16] Lurie, *Louis Agassiz,* 379–381.

[17] A. H. Dupree, *Science in the Federal Government. A History of Policies and Activities up to 1940* (Cambridge, Mass.: Harvard [Belknap] Press, 1957).

opened in 1888,[18] one year after the opening of the Misaki Marine Biological Station by the Imperial University of Tokyo.[19]

The fact that Dohrn's ideas of administering marine stations had predecessors in the minds of those who had earlier tried to open sea-side laboratories does not detract one whit from his accomplishment. Dohrn's plan was for much more substantial institutions. And it was he who translated the idea into magnificent reality: he developed an imaginative and practical plan for financial support, and an equally imaginative and practical plan for the physical and technical organization of the laboratories.

Dohrns and von Baers as European Families

In 1869, just a year after Anton Dohrn started to think about marine stations, he wrote to von Baer to enlist his help in procuring financial aid in Russia for the purpose (letter #4 in this volume). Von Baer was then living in Dorpat, having retired in 1867, as stated above, from active membership in the Imperial Academy of Sciences in St. Petersburg, where he had worked for 34 years, not only as a remarkable scientific investigator but also as a highly respected scientific statesman.

Thirty-six of the 41 letters that constitute the main body of the collection that we present here convey messages that passed between Anton Dohrn and von Baer. The others, among the five which begin this volume, elucidate to some degree the relationship between the Dohrn and the von Baer families and thus shed some background light on our two protagonists.

Anton Dohrn's blossoming as a man of eminent distinction was hardly an eccentric development of a sport on an otherwise symmetrical family tree. His paternal grandfather, Heinrich Dohrn (1792–1852), born a Pomeranian, the son of a surgeon, became an industrialist in Stettin; he was the co-founder of a sugar refinery there in 1817. Perhaps of equal interest to him and to his descendants were his extraordinary musical ability and interest. He was a talented violinist, and his home was a highly influential musical center in Stettin.

Anton Dohrn's father, Carl August Dohrn (1806–1892), was a natural son of Heinrich; he was legitimatized in 1814. His to us perhaps seemingly unconventional birth (not so unusual, however, in the Germany of his class and his time) was followed by a highly individualistic pattern of life. From childhood on he was fortunate to live in comfortable circumstances, and thus was never obligated to choose to follow such paths as might be dictated by urgent needs for funds for himself or his family. He was free to vacillate, and so he did. If you were to look him up in a current volume of the *Neue Deutsche Biographie,* you would find him described

---

[18] F. R. Lillie, *The Woods Hole Marine Biological Laboratory* (Chicago: University of Chicago Press, 1944).

[19] H. Terayama, "The Misaki Marine Biological Station (Faculty of Science, University of Tokyo)," *Internat. Soc. Dev. Biols. Newsletter,* May 1979 [no vol. #]: 3–4.

as an "entomologist and man of letters" [*Literat*].[20] While the entry did point out that he had interest in a sugar refinery, it first specified that he was highly musical. (His musical genes, presumably inherited from his father, were indeed potent; they were still strongly expressed in his daughter Anna's grandson, Wilhelm Furtwängler.)

His life followed a rather zigzag path. Entomology was an early interest. He sang; he enrolled in the Law Faculty in Berlin. There he circulated in high society, in a group numbering distinguished men of letters and musicians, among them Felix Mendelssohn-Bartholdy. He traveled; he translated Spanish drama. And he was a successful industrialist. If he was a sort of professional dilettante, he was, as Theodor Heuss put it, a dilettante in the grand style; Heuss considered him "naively egocentric."[21]

Among his other vagaries, he quarreled with his father over his desire to marry a particular divorcée. The dispute was ultimately brought to resolution through the intermediation of Alexander von Humboldt, whom Carl August had met in Berlin in 1832 through Mendelssohn. He married her only later, in 1837. She was the mother of his natural daughter Anna (1832–1892) and of their three legitimate sons, Heinrich (1838–1913), Wilhelm (1839–1903), and Felix Anton, named for his godfather Felix Mendelssohn but usually called by his middle name.

So much for a minimal outline of a few aspects of the lives of some of the Dohrns who preceded Anton. Karl Ernst von Baer's antecedents were seemingly less affluent, and seemingly less commercially inclined, but they too moved in high circles. Members of the Russian aristocracy, they were not especially elevated within its ranks. Karl Ernst's forebears, like those of Anton Dohrn, had been originally Prussian; his great-great-great-grandfather on his paternal side emigrated from Westphalia to the Baltic lands in 1597. What we have recently been calling the Baltic Republics were in Karl Ernst's time Russian provinces (in Czarist days the Russians called their administrative subdivisions by the French word *gouvernements*). Their borders and names and allegiances have changed from time to time; the Baers first settled in what was then Livonia, which no longer existed under that name in 1792 when Karl Ernst was born on the estate Piep near Jerwen, in Estonia.

Karl Ernst was a fervently patriotic Estonian. He knew the vernacular Estonian language, and his first publication, as a student, was a review of a short Estonian manual for midwives,[22] but he wrote exclusively in German throughout his life. His early education was in Estonia. His family had wished him to spend his university years in Germany; his father had been trained in law in Erlangen. Karl Ernst insisted upon attending the university in Dorpat; his father permitted it only on condition that he learn the Russian language. He met this requirement tech-

[20] *Neue deutsche Biographie,* 1959, 4:56 (article Carl August Dohrn; no author named).
[21] Heuss, *Anton Dohrn,* 46.
[22] For details see Raikov, *Karl Ernst von Baer,* 428.

nically, but never felt sufficiently confident in his command of it to write it; the small minority of his publications that later appeared in Russian were translated by others. When he lectured in Russia he did so in Latin, then the universal language of scholars.

He received his medical degree in Dorpat in 1814, then left Estonia to spend the main part of his professional life elsewhere, returning to Dorpat to live only in 1867, and remaining there until he died in 1876. The family estate in Piep was until the early 1830s the home of his elder brother Ludwig. When the latter died childless in 1834 Karl Ernst took over its management to preserve it for his family, but did not himself take up residence there. He lived and worked for 17 years in Königsberg (1817–1834), and next for 34 years in St. Petersburg (1834–1868). It was during his subsequent retirement years in Dorpat that he wrote all the letters reproduced here.

There seems to have emerged from the available records no evidence to indicate that this family lived in luxury; it says something about its station in life that a number of von Baer's forebears entered professional military careers, as his uncle Major Karl Heinrich von Baer had hoped that Karl Ernst might do[23]; Karl Heinrich and his wife, childless, raised Karl Ernst during his first seven years. Karl Ernst, in order to subsidize his study in medical school, was obliged to borrow funds from a brother. But the Estonian school he attended as a boy educated sons of the nobility, and during a number of his years in St. Petersburg he was a protégé of Grand Duchess Elena Pavlovna, wife of Grand Duke Michael, a brother of Czar Nicholas I. She was greatly interested in science, and chose von Baer to tutor her two daughters in it. In this respect, von Baer's life in society might be considered to have attained the same level as that of Carl August Dohrn, who in his Berlin days consorted socially with Friedrich Wilhelm IV, and that of Anton Dohrn, who became a confidant of Kaiser Wilhelm II.

If there was no artistic or musical equivalent of Mendelssohn in the entourage of von Baer, Karl Ernst was otherwise well connected through family acquaintanceships. Count Alexander von Keyserling[24] was a close

---

[23] Von Baer, as an explorer at sea, in due time had connections with the Russian Navy. His professional connections with Admiral Lütke and with Admiral von Krusenstern are taken up in the text. His son August Emmerich (1824–1891) became a general major in the Russian Navy.

Karl Ernst had three sons and one daughter in addition to August: Karl Friedrich Julius (1822–1843), Alexander (1826–1914), Marie (1828–1900), and Hermann (1829–1866). Marie married Karl Magnus von Lingen, M.D. (1817–1896) on 24 February 1850.

Recently a list of all von Baer's descendants, male and female, has been published as follows: V. Kaavere, *Die Nachkommen Karl Ernst von Baer* (Tartu: Estonische Akademie der Wissenschaften, Institut für Zoologie und Botanik, 1990). It lists Karl Ernst's first child as Magnus Ludwig Conrad (1820–1828). Kaavere identifies in detail 242 individuals, born between 1820 and 1988, and provides many fine photographs. The dates I have given for von Baer's children are those presented by Raikov and many others; Kaavere points out the existence of two letters suggesting that 1830 is the correct date for Hermann's birth.

[24] Graf Alexander von Keyserling was a geologist and palaeontologist. For some of his activities and relationships with von Baer, see his daughter's memoirs of him: H. Freifrau

friend, and von Baer enjoyed a number of associations with Alexander von Humboldt,[25] beginning in 1836. When in that year the Paris Academy of Sciences awarded von Baer a medal for his contributions to embryology, it was transmitted to him by von Humboldt. Von Baer called von Humboldt "the pride of German science" in his inscription to him in the presentation copy of his monograph on *Entwickelungsgeschichte* (now on the shelves of the Library of the Stazione Zoologica). Among others of the social and scientific elite who befriended von Baer was Admiral Ivan Feodorovich von Krusenstern. When, in 1819, as a young man before marriage, von Baer had thought of exploring the islands of Novaya Zemlya, the possibility was arranged for him by von Krusenstern, who had been the first Russian global circumnavigator[26]; von Baer's trip, however, was postponed because of his marriage in 1820. When in 1834 von Baer moved to St. Petersburg to find that there was no residential space available for him in the Academy at that time, von Krusenstern arranged housing for him in Navy living quarters. Von Baer would later, as we shall see, have other meaningful professional connections with such influential naval figures.

In sum, von Baer and his family may have been even more remote from the bourgeois than the Dohrns; for them there was no hint of commercialism, no trace of the mercantile. Living their lives apart from academia, neither Dohrns nor von Baers were tainted by pedantry, nor did they ever become arrogant as a result of their friendships among the high and mighty. Admittedly they moved easily among the elite, but they rose to these high levels as a result of their intellects and not through social conniving.

The Scientific Interests of the Two Protagonists

As it transpired, it was not through such social frivolities as acquaintanceships in the great courts of Europe that members of the two families who interest us developed their mutual friendship. It was science that was the basis for that. When Carl August Dohrn wrote to von Baer the letter that opens this collection (#1, 10.9.62), it was in response to a request that von Baer had made to him for a sample of Baltic sea water. Although science was to be a continuing basis for the correspondence, the letters are not stiffly professional in tone. Carl August's letter about the Baltic sea water mentions that von Baer was acquainted with his zoological son Heinrich (there was no reason then to mention Anton). It was not a mere reply concerning sea water; it was also a friendly personal

---

Taube von der Issen, *Graf Alexander Keyserling. Ein Lebensbild aus seinen Briefen und Tagebüchern*, 2 vols. (Berlin: Georg Reimer, 1902), passim.

25 F. W. H. A. von Humboldt quoted in *Cosmos*, 1858, 4:42 some of von Baer's data on *Bodeneis*.

26 Between 1803 and 1849, there were thirty-six Russian voyages around the world, according to M. Mitchell. *The Maritime History of Russia 848–1948* (London: Sidgwick and Jackson, 1949), 18.

greeting. This is true of all the letters presented here. Our single letter from Heinrich Dohrn to von Baer (#2, 3.6.63) also makes a scientific comment, it seems to reply to an earlier question from von Baer about some sea shells which von Baer collected; it concludes in almost bantering terms. And our third letter here, signed by Carl August as President of the *Stettiner Entomologische Verein*, addressed to von Baer as a member of the *Verein*, congratulates "Ursa major" on the 50-year Jubilee of his doctorate (#3, 6.11.64). (Von Baer had shortly before, in 1859, been one of the founders of the Russian Entomological Society and was its President in 1860 when it first met; insects had been the subjects of occasional papers by von Baer since 1829.) All our letters concern matters of scientific import, but they are all couched in easy terms of personal amity. It is all very well for biographers to comment that the stations in life of members of these two families were comparable, but the friendliness of their exchanges of ideas, as expressed in their own words, exemplifies the unique immediacy that we have pointed out.

Perhaps it is appropriate here to mention some of the scientific interests of our protagonists, as representatives of both families, so that we may understand how their activities and interests happened eventually to overlap in the enterprise involving the Naples Stazione.

One of Anton's earliest interests was in entomology, resulting no doubt at least partly from exposure to it in his home through his father's involvement in it. Carl August may have been technically an amateur as an entomologist, but his accomplishment in it attained professional standards. He made a collection of beetles that was famous; he was instrumental in the establishment in 1840 of an important entomological journal, the *Entomologische Zeitung*, and edited it until 1887, when his son Heinrich succeeded him as editor. Heinrich was a truly professional scientist. He was trained in natural sciences at Bonn and Zürich; he set up Stettin's first museums. He had perhaps fewer strings to his bow than his father had had before him, but he did have a double career: he was a parliamentarian as well as a scientist. Elected to the Reichstag in 1875, his interests were deeply political.

Anton's early entomological interest was not in beetles, as his father's had been, but in the true bugs, the Hemipterans. His first publication on them appeared in his father's journal in 1858, when he was only 18 years old. He early thought of becoming a zoologist, attending the Universities of Königsberg, Bonn, Berlin, and Jena, but he became disenchanted, and for a while thought of becoming instead a historian, or perhaps a book-dealer. His studies were briefly interrupted by a military interlude. He was reconverted to zoology by the spread of Darwin's ideas to Germany, to which he was exposed quite promptly as a result of having become acquainted with Ernst Heinrich Haeckel in Jena in 1862. Dohrn grasped immediately the relevance of evolution theory for taxonomy and for anatomy and embryology, and returned to concentrate his central scientific thought on biology. He continued his strictly scientific studies throughout his life, never completely diverted from them by the entrepreneurial and administrative pursuits related to his founding of the Stazione and his guiding of its activities.

The virtuosity and versatility of members of three generations of Dohrns have earlier claimed our attention; science was only one of their interests. Von Baer's interests ranged more narrowly outside science, but far more widely within it.

In his early childhood von Baer was fascinated by the plant world. Later trained in medicine, the only time he devoted to medical practice was during the cholera years (1831 to 1832 in Königsberg). During a short period of post-medical *Lehrjahre* (around two years in Würzburg, Vienna, and Berlin) he became acquainted with embryology; this was the field in which he was to make his most familiar contribution. He spent his first teaching years (1817–1834) in Königsberg. There he discovered the true mammalian egg, in the ovaries of dogs and women, correcting for all time longstanding errors of interpretation.[27] Having learned in Würzburg how to handle young developing chick embryos, he now studied them in detail, and these investigations and his profound interpretations of their results demonstrated conclusively that development is epigenetic, that is, that differentiation is progressive rather than the mere unfolding of something preformed that is already there; it is true change that dominates embryonic development. He also suggested that much of embryonic change is accomplished by mechanisms that are comparable in all embryos, mechanisms involving the forming and folding and fusing of layers. He thus set a stage that has been occupied by all embryologists who followed him, a stage upon which embryologists continue to act out their roles today. Von Baer published the first volume of his major embryological monograph, *Ueber die Entwickelungsgeschichte der Thiere. Beobachtung und Reflexion*, in 1828. The second volume appeared, incomplete, in 1837, its proofs uncorrected, after von Baer was ensconced in Russia.[28] It was surely in large part because of his esteem for von Baer as an embryologist that Dohrn was later to seek his support for the establishment of a marine station; it was on embryology as an experimental science that the most important work was to be concentrated during the earliest years of the station.

After von Baer moved to St. Petersburg embryology was no longer his major pursuit. His first appointment there was as Librarian of the Foreign Division of the Imperial Academy. It was only in 1846 that he was formally appointed Academician for Comparative Anatomy and Physiology. That is not to suggest that he spent the intervening twelve years concentrating on books, before again devoting his hands, eyes, and mind to the more active work of science.

---

[27] K. E. von Baer, *De ovi mammalium et hominis genesi epistolam ad Academiam Imperialem Scientiarum Petropolitanam* (Leipzig: Leopold Voss, 1827).

[28] K. E. von Baer, *Ueber Entwickelungsgeschichte der Thiere. Beobachtung und Reflexion.* (Königsberg: Bornträger), 2 vols. [1828, 1837]; the third volume originally drafted as conclusion to vol. 2, was published posthumously in 1888, ed. L. Stieda (Königsberg: Koch).

During his Russian years, he made a number of contributions to a wide variety of scientific enterprises. For some of these his interests were longstanding and reached back into his pre-Russian life. It will be remembered that as a youth he had thought of traveling to Novaya Zemlya; he finally explored there in 1837, the first naturalist to collect specimens there. He traveled widely in the Russian north and throughout many parts of Europe and Scandinavia. He made significant discoveries in geography and also in various areas of investigation that we would consider to be aspects of ecology. He thought of organisms, including man, as influenced by and as influencing their environment in mutual interrelationships.

His interests in man were focused at various levels. Physical anthropology became one of the major areas of his efforts. His first book, based on lectures he delivered in Königsberg (1824), had been on this subject.[29] In St. Petersburg, he established an important craniological collection; as a result of his interest in it, Russia was one of the first countries in which physical anthropology became established as a separate discipline.[30] Later he convened an international meeting of anthropologists in Göttingen that resulted in the founding of the German *Archiv für Anthropologie.* Ethnography, and the possible interrelationships of the races of man, also commanded his attention; such considerations were furthered during his geographical explorations and were enhanced by his observations during his wide travels.

How did he integrate within his own mind his diverse interests? By envisioning all of nature as changing and developing. *Werden* was his key to unity. *Bildung* was for him *Umbildung; Entwicklung* was the unifying feature in cosmos and man and all between.

Although his formal title changed from time to time in the Academy, he was always revered there as a major statesman of science and as one of its most active practitioners. In fact, he was enormously respected not only in St. Petersburg, but also in the wider world, for his whole *oeuvre,* and for his broad and deep philosophical reflections, and Dohrn would have had such distinctions, as well as his contributions to embryology, in mind when he enlisted his aid.

The Content of the Letters

The earliest of the letters to von Baer presented here are included in our publication to help illuminate the background against which the following ones may be regarded. The main import of the messages in the letters from 1869 on relates to the establishment of the Stazione Zoologica in Naples. But some of the letters also deal specifically with particular

---

[29] K. E. von Baer, *Vorlesungen über Anthropologie, für den Selbstunterricht. Erster Theil* (Königsberg: Bornträger, 1824).

[30] A. Vucinich, *Science in Russian Culture. 1861–1917* (Stanford: Stanford University Press, 1970), 225–226.

scientific thoughts of the two eminent biologists who exchanged them. It was a distinction of both individuals that their involvement in the promotion of scientific affairs at an advanced administrative level did not diminish their strictly scientific accomplishments in basic biology proper. We shall discuss first the affairs of the Stazione, then science itself.

Establishment and support of the Stazione Zoologica

We began by saying that the idea of founding one or more marine stations first entered Dohrn's mind in 1868. In 1869 he wrote to von Baer to request his help for such a venture (#4, 30.12.69). What Dohrn was asking at this time was mainly moral support, a written statement from von Baer saying that the plan had merit, and that Dohrn would have the capacity to see it through. Von Baer seems not to have replied to the letter. In 1871, Dohrn's father Carl August wrote to von Baer about the Stazione, but without making any specific overt request for aid (#5, 2.10.71). In 1873, however, Anton himself again wrote to von Baer, again appealing for help (#6, 2.1.73); this time he expressed his hope that von Baer might aid in procuring support for the enterprise in the form of subvention from the Russian government. This was a letter from one embryologist to another, and von Baer answered it as an embryologist. But before returning to the content and significance of von Baer's reply, let us examine for a moment Dohrn's reasons for needing help at all. In short, they were financial.

Even if land near the sea might be available *gratis*, as proved the case in Naples, a building would have to be designed and erected. The edifice, in Dohrn's dream, would require more than walls. Aquaria with circulating fresh and sea water would have to be supplied for each workroom, and in each work space there would have to be outlets for all types of available utility. The laboratory would require a boat well equipped with collecting gear for procuring living specimens; not only sailors but highly skilled collectors would be needed to man it. (Highly skilled indeed they were in Naples; when I worked on blenny and goby eggs at the Stazione, I required to have them at an early gastrula stage, when the transparent yolk, less than 1 mm in diameter, was half-covered by a diaphanous membrane, difficult to see even with a microscope; these were supplied to me routinely.) The station would supply investigators not only with aquaria and living specimens but also with all the material appurtenances they might require for their work: worktables, glassware and other disposable supplies, chemical reagents, all the usual laboratory apparatus of the day such as ovens, incubators, sterilizing equipment, microtomes and their knives, and so forth—everything except microscopes. In a laboratory planned by a German, *Diener*s would be abundant, and helpful personnel of many ranks. Living quarters might be provided for some, in the European fashion. Last but not least, the availability of books and periodicals in a well-stocked library would be a prime necessity. When the Naples Stazione began to function according to such a pattern, it became the first of all independent research institutes, and the model for all to follow; Simon Flexner visited it before setting up the Rockefeller Institute. (What impressed Flexner most about the Stazione seems to have been the independence and freedom permitted to its workers: "Their results are their own," he is reported to have written to Christian Herter in 1903. "Everything is supplied them for their problems, but their results are their own. Any discoveries they make are theirs; and blunders theirs too."[31])

The idea behind the establishment of such a research institution was the brainchild of Anton Dohrn alone. Once the original idea took on its own life, Dohrn was faced with the challenge of finding financial resources for starting and maintaining his new type of laboratory, and he did this with equal originality.

The story of Anton Dohrn's ultimate success, after early difficulties, in obtaining premises for a laboratory in Naples's most beautiful park has been told elsewhere and need not be repeated here. Many details of his struggles to obtain funding are also published elsewhere, but some of these may be briefly summarized again since it was in connection with these that Anton Dohrn first wrote to von Baer.

Dohrn at the outset had hoped that his own funds would suffice, then soon thought that additions to them might be made by his father, as proprietor of a highly successful commercial venture who lived very comfortably. Carl August Dohrn at first refused to cooperate, believing the whole concept to be merely a fanciful vagary. Ultimately, however, he relented, and the enterprise went forward with his support, not only financial but also moral, as his letter of 2.10.1871 to von Baer indicates (#6). However, the Stazione's maintenance, in the style and on the scale envisioned by Anton, promised to be, and did become, extremely demanding financially. The building in Naples was projected to incorporate an aquarium for displaying living marine creatures to a curious public; Dohrn had hoped that the admission fees to the aquarium would provide funds sufficient to support research. The aquarium was indeed included in the building (and is still popular among Neapolitans as well as among tourists) and has steadily produced income, but of course not enough to subsidize research.

Another contribution made by the Stazione itself toward its own support resulted from the work of its Conservation Department under the exceptionally distinguished guidance of Salvatore Lobianco. As the virtues of the Stazione and the skills of its collectors in obtaining and maintaining marine organisms alive became known abroad, requests were received for biological specimens to be shipped away for use elsewhere. These were provided, sometimes living, more often preserved. Placing a bony fish into alcohol or formaldehyde or phenol, or some such preservative, is not always a hard task once the fish has been located and caught. But the challenges of preserving delicate and fragile marine in-

---

[31] Simon Flexner's letter to Christian Herter is quoted in C. Groeben, "Anton Dohrn—the Statesman of Darwinism," *Biol. Bull.* 168 (suppl.) (1985): 4–25; quotation from p. 22.

vertebrates are far more difficult to meet. The Conservation Department developed splendid techniques for the preservation of invertebrates as well as vertebrates, and distributed preserved specimens widely. This was an important contribution to promoting the advancement of marine biology at a time when, especially after the expedition of the *Challenger* (1872–1876), the rich variety of marine creatures was beginning to be appreciated; fortunately it was a practice lucrative for the Stazione.

The Stazione also earned funds for its support by itself introducing new ventures in the guise of publications. In 1880 it began to edit and publish an annual series of volumes entitled *Zoologische Jahresberichte,* a publication distinctive in several ways. Its coverage was truly international, and it was the first review journal to cover all fields of zoology. It became an important factor in stimulating the development of zoology at the time that this was becoming organized as a separate distinct discipline. Also, in 1879 the Stazione began to publish a periodical for reporting the results of the investigations carried out in its laboratories; this was widely subscribed to and thus important in the furtherance of the new disciplines, such as physiology and experimental embryology, which were being so greatly advanced by the investigators performing their experiments in the Stazione. Finally, the year 1880 saw the start of a new monograph series, *Fauna und Flora des Golfes von Neapel und der angrenzenden Meeresabschnitte;* this has so far numbered 40 volumes, in large format, all consummately accurate in taxonomic and morphological detail and sumptuously illustrated. This ambitious program of publication not only strengthened zoology and marine biology, but also produced monetary profits that benefited the Stazione itself.

But still, all such income, even in the aggregate, could not without further amplification suffice to meet the expenses of research support. Fortunately, in 1873 Anton Dohrn had the brilliant inspiration of having the Stazione's work places subsidized by major official entities or organizations; the idea came to his mind as he contemplated the fact that hospital "beds" were being subsidized by corporations, or even by affluent individuals, who guaranteed to hospitals annual contributions, or who added substantially to their endowment, on the condition that the donors could name the patients to occupy the beds they sponsored. Dohrn supposed that government bodies, or Academies, or universities or other organizations, might similarly support worktables at the Stazione. This proved to be a feasible and very happy solution to the problem of major financial support.

Against this background of information concerning some of the financial aspects of establishing the Stazione, let us return to the two letters from Anton Dohrn that we have already mentioned (#4 and #6) as having requested aid from von Baer. The first governmental regime to promise support of one or more work places had been that of Berlin, in 1873, through its Academy of Sciences. Within the same year, ten work places in all were assured of support; Russia was included among the nations promising help.

Von Baer did not reply to Anton Dohrn's first letter, that of 30.12.1869, and he explained why when he replied promptly, on 15/27 January 1873 (#7) to Dohrn's second letter, that of 2.1.1873 (#6). He said that when he received the first of the two letters, he had then been uncertain that Dohrn's plan could be effectively carried through, but that subsequently, even before receiving Dohrn's second letter, he had heard fine reports of progress of the enterprise, and he now was indeed willing to help. Having been appealed to as an embryologist, his reply, as we have said, was that of an embryologist; specifically, he wrote that he had some years ago traveled to Trieste to attempt some studies on sea urchin eggs, but had gone there at the wrong time of year, since he had no way to ascertain when the spawning season occurred. (What he did not say to Dohrn was that even when he did have eggs to study in Trieste, he was unable to complete his work on them since they were thrown out by the chamber maid who cleaned his hotel room which was his sole work place.) In agreeing to provide help, von Baer offered to write a public plea for support in the German-language *St. Petersburgische Zeitung*; dated 9 August 1873, this appeared in print,[32] and it presented further details of his misadventures in Trieste.

As readers familiar with Russian history will know, bureaucracy did not have to await the Revolution to enter into Russian halls of power. (Von Baer once commented that he had enjoyed being in Novaya Zemlya when it was unpopulated because then there were no officials around.) Representatives of the Russian Empire may indeed have been prompt in 1873 in making promises to Dohrn concerning support of the Stazione, but while Russia supported its first table as of 1.4.74, there was delay, exasperating to Dohrn, in confirming the support of a second. This is where von Baer was approached and where he took action. A number of the letters in our collection that passed between Dohrn and von Baer referred to the delay and to attempts to ameliorate the situation.

In a letter to Dohrn of 18 August 1873 (#9) von Baer reported that he had recently been in St. Petersburg and that he had spoken to the Secretary of the Academy and had received his agreement that two tables would be supported. Later in the month (#11, 31.8/12.9) he wrote to Dohrn that Count Dmitri Andreevitch Tolstoi, the Minister of Education, had submitted such a plan to the Academy. Further efforts by von Baer to speed up the implementation of Russian support are referred to in letters 15, 17, and 18 below. Letter 17 is of special interest since in it von Baer promised to write to the Secretary of the Academy; he did so, and that letter is published in Lukina's edition of von Baer's *Correspondence*, and

---

[32] K. E. von Baer, "Die Zoologische Station in Neapel," *St. Petersburger Zeitung*, No. 219 (1873). [We do not have access to original copies of this journal. References to various articles by von Baer that appeared in it do not carry page numbers in von Baer's own list of his writings in his autobiography, nor in the references to them in the comprehensive bibliographies of his publications presented by Stieda, Raikov, or Sutt nor in references to them by Lukina.]

will again be referred to below on this page, together with others now in St. Petersburg.

All the letters here about to be enumerated as published by Lukina in Leningrad confirm von Baer's interested involvement in supporting in Russia the cause of the Stazione. But the assistance in Russia of Dohrn's future father-in-law, Georg von Baranowski (Dohrn married Maria von Baranowska in 1874), was also helpful and important in the promotion of the Stazione in Russia. Von Baranowski, former Governor of two Russian provinces, was highly placed in government circles; it was through his good graces that when Anton Dohrn spent three weeks in St. Petersburg in 1874 he had access to the Finance Minister von Reutern and to Count Tolstoi, and also to the German Ambassador to Russia, Heinrich Prince Reuss.

It was von Baer, however, who was in the position to speak to scientists for the scientific importance of the Stazione's work, and he could address them in the authoritative voice of a highly influential member of the Academy itself. Several documents not included in the collection that we present here attest to his direct communication with the Academy.

A letter survives in St. Petersburg, dated 29.1.74, that von Baer wrote to K. S. Veselovsky, Permanent Secretary of the Academy from 1857 to 1890; this is the letter that he promised Dohrn to write in our letter #17.[33] It specified that Dohrn had not been able to ascertain from Count Tolstoi that Russia would support a second table. Von Baer pointed out, in this letter, what important contributions were currently being made by Russian nationals to embryology–a thoroughly just evaluation: we think of C. F. Wolff, H. C. Pander, N. Kleinenberg, E. Metchnikoff, A. von Kowalewski, and others.

He again wrote to Veselovsky about the Naples predicament on 23 April 1874[34] and on 27 May 1874.[35] He had previously, on 19 October 1873, written about the Stazione to a physician, Karl Renard,[36] who at that time was Secretary of the Moscow Society of Naturalists; later he would become its Vice President and then its President. In spring 1874 when von Baer was writing to Veselovsky, he also wrote a letter promoting the cause of the Stazione to Admiral Feodor Lütke, the President of the Academy.[37] Lütke and von Baer shared many interests.

Lütke had been the first to spark development of Russian craniology when in 1830/1831 he contributed to the Academy skulls that he had collected during a voyage around the world in 1826–1829. Thus, beginning the collection that was to lead to the development of the craniological

---

[33] K. E. von Baer to K. S. Veselovsky, Dorpat 29 January 1874, in Lukina, *Perepiska Karla Bera* II: 222.

[34] K. E. von Baer to K. S. Veselovsky. Dorpat 23 April 1874, ibid. II:225–226.

[35] K. E. von Baer to K. S. Veselovsky, Dorpat 27 May 1874, ibid. II:227–228.

[36] K. E. von Baer to K. I. Renard, Dorpat 19 October 1873, ibid. IV:111–112.

[37] K. E. von Baer to F. P. Lütke, Dorpat 27 March 1874, ibid. I:154–155.

museum later expanded and organized by von Baer, he initiated the moves that led to the development of physical anthropology in Russia. Lütke, like von Baer, had visited Novaya Zemlya, where he had been the first to map its coasts, and had also traveled to Russian America (later to become Alaska), territory in which von Baer was much interested, and about which he would write.[38] Lütke was also instrumental in founding the Russian Geographical Society, an enterprise in which von Baer was also involved. Grand Duke Konstantine was its first official President, and Lütke was its Vice President for 20 years. Von Baer, when the Society held its first meeting in 1845, presided over, and addressed a lecture to, its section on ethnography. Lütke was also actively involved in planning and organizing the great celebration held in St. Petersburg in 1864 to celebrate the Golden Jubilee of von Baer's doctorate—surely in part a result of long friendship, but there was also an official dimension to his participation, since Lütke was then President of the Academy, a post he held from 1864 until his death in 1882. Lütke personally contributed over 400 rubles toward a prize named for von Baer, first announced at the celebration.

Ultimately two tables were financed by Russia in Naples, the second from 1.1.75 on. Also, the Russian Imperial Navy supported a third Russian table, from 1886 to 1893, sending young officers to train in preparation for their own later expeditions; the Russian Navy also sent a number of representatives to learn the Stazione's methods of preservation. It is tempting to speculate, in the total absence of any supporting data, that Admiral Lütke might have been interested in such activities had he known of them, but he died in 1882.

In fact, it was certainly long before that the earliest *rapprochement* had been made between scientists and naval officers. We remember (see p. 5) that Carl Vogt had failed in an attempt to cooperate with the French Navy in the 1840s.

Eventually it was on the initiative of the Italian Navy itself that the first approach on behalf of a ship-of-the-line was made to the Naples Stazione. Around 1878 an Italian naval officer asked Dohrn's advice regarding methods of collecting marine plants and animals during the coming circumnavigation of the corvette *Caracciolo*; Dohrn suggested that naval officers be trained at the Stazione in techniques of collecting, sampling, classifying, and preserving. Out of such beginnings grew a more intimate collaboration. In 1881 Lieutenant Gaetano Chierchia spent several months training at the Stazione before becoming second officer in command of the *Vettor Pisani* circumnavigation. The corvette sailed from Naples on 20 April 1882 and when she returned on 29 April 1885 it was with a collection of 1600 specimens.[39] Thus while the Russian

---

[38] F. P. Wrangell, *Russian America. Statistical and Ethnographic Information.* With additional material by Karl-Ernst Baer. Trans. from German ed. of 1839 by Mary Sadouski. (Kingston, Ontario: Limestone Press, 1980).

[39] C. Groeben, "The *Vettor Pisani* Circumnavigation (1882–1885)." *Deutsche Hydrographische Zeitschrift*, Ergänzungsheft B 22 (1990): 220–234.

Navy may have been early in providing continuing financial support for the Stazione, its officers were not the first to take advantage of its facilities and advice.

Some scientific interchanges in the letters

We have from time to time in these pages been referring to powers in a more or less political sense, now we may conclude by focusing on the power of ideas. We emphasized at the start that this correspondence was an interchange between scientists, and that its subject was primarily science. What we have said about it, however, has concentrated upon primarily administrative aspects of science, and less upon science as an intellectual function. Thus it remains, in conclusion, to discuss some of the ideas of our protagonists as expressed in these letters, especially as they related to other contemporary views.

It is evident that both Dohrn and von Baer shared an interest in insects and in entomology, but for von Baer this was never a major preoccupation. There are so many insects of so many varieties and they are so ubiquitous that it is hardly a distinction for biologists to be fascinated by them; what would be surprising would be a lack of wonderment about them. It is also clear, as we have already said, that Dohrn addressed von Baer in the letters as one embryologist to another, thus that others of their biological interests besides entomology overlapped. They were both interested, in very different ways, in aspects of what we now call evolution.

Attention was drawn on pp. 2–3 to the fact that Dohrn's once failed interest in zoology was reawakened when he became acquainted with Darwin's ideas. Charles Darwin was born on Abraham Lincoln's birthday on 12 February 1809; the *Origin of Species* was first published in 1859; it appeared in Bronn's German translation in 1860. Haeckel began to read it promptly, and he lectured on it in Stettin in 1863.

Haeckel also lectured on Darwin to university students in Jena in the winter semester of 1863, and student notes taken at these lectures formed the basis of Haeckel's second book, *Generelle Morphologie.*[40] (His first book, published in 1862, had been a scientific monograph on the Radiolarians[41]; he also had meantime published several articles in periodicals.) *Generelle Morphologie,* like all that Haeckel had so far published, was addressed to a professional readership. Its character is revealed by its overweening subtitle: *Allgemeine Grundzüge der organischen Formen-Wissenschaft, begründet mechanisch durch die von Charles Darwin reformierte*

[40] E. H. Haeckel, *Generelle Morphologie der Organismen. Allgemeine Grundzüge der organischen Formen-Wissenschaft, mechanisch begründet durch die von Charles Darwin reformierte Descendenz-Theorie.* Bd. 1: *Allgemeine Anatomie der Organismen. Kritische Grundzüge der mechanischen Wissenschaft von den entwickelten Formen der Organismen, begründet durch die Descendenz-Theorie.* Bd. 2: *Allgemeine Entwicklungsgeschichte der Organismen. Kritische Grundzüge der mechanischen Wissenschaft von den entstehenden Formen der Organismen, begründet durch die Descendenz-Theorie* (Berlin: Georg Reimer, 1866).

[41] E. H. Haeckel, *Die Radiolarien (Rhizopoda radiria). Eine Monographie* (Berlin: Georg Reimer, 1862).

Descendenz-Theorie. Erster Band. Allgemeine Anatomie der Organismen. Kritische Grundzüge der mechanischen Wissenschaft von den entwickelten Formen der Organismen, begründet durch die Descendenz-Theorie. . . . Zweiter Band: Allgemeine Entwicklungsgeschichte der Organismen. Kritische Grundzüge der mechanischen Wissenschaft von den entstehenden Formen der Organismen, begründet durch die Descendenz-Theorie. It was not at all well received by its professional readers. Unlike Haeckel's later books on the same subjects, which passed through many editions and were translated into many other languages, it was never reissued in Germany or elsewhere. It was excessively wordy, and characterized by many exaggerations and over-systematizations. Dohrn however could see through the froth to the central core, especially to the significance of Haeckel's emphasis on relationships between the development of embryos to the descent of their parent stock. He thought well of the book and was profoundly influenced by what he learned from it. Both when the book first appeared and later, Dohrn was strongly out of sympathy with many of Haeckel's philosophical opinions, but when he read *Generelle Morphologie* he concentrated on its implications for organisms, and he said in 1897 that in the "book lay the foundations of a new science."[42] One of Haeckel's excesses was his emphasis on what was later to be called the Biogenetic Law, which stated that embryos resemble their ancestors. Thus, he concluded, by studying embryos one could ascertain what were their ancestral species. This was not a new idea on Haeckel's part. His role was to formulate around it an incontrovertible doctrine. It had been refuted even before Haeckel's emphasis on it by none other than von Baer, who had pointed out in 1828[43] that embryos resembled not adults but only other embryos. Nonetheless, the Biogenetic Law as framed by Haeckel had a great vogue, and Anton Dohrn was only one of many to be inspired by it.

Dohrn's early scientific work concentrated first on insects, later on arthropods more generally. Insects and some other arthropods metamorphose through distinct larval stages. When Dohrn went to Messina in 1868 he was accordingly especially fascinated by the crustacean larvae he saw there.

Many marine larvae float near the surface as a means of dispersal, especially if their adults are sessile or are otherwise not highly mobile. Part of the great aesthetic beauty of the larvae results from the elaboration of the fragile extensions of their bodies transformed to floating organs. While observations of extremely small organisms collected at the surfaces of the sea or near to its shores date back to the eighteenth century, details of the structure and form of these delicate creatures had to await the improvements to microscopy brought by the nineteenth century. Also, Johannes Müller's work greatly enhanced interest in such larvae after 1845 when he showed for the first time that echinoderm

---

[42] Kühn, *Anton Dohrn*, 23.

[43] Von Baer, *Ueber Entwickelungsgeschichte der Thiere*, I:220.

plutei metamorphose into brittle stars[44]; until then they had been considered separate species. The desire to study the development of metamorphosing marine larvae into known adult forms pointed straight to the advantages of aquaria that could support their development, and when Dohrn contemplated Darwin's ideas he could see new values in being able to collect and maintain such larvae.

We now take for granted that marine larvae are highly numerous and important constituents of what we call the plankton. In the 1860s the word plankton did not yet exist, nor was there an idea that such an entity existed that might be designated by a name. The word was not coined until 1887, and the first expedition planned expressly for its collection and study did not sail until 1889,[45] 27 years after the first departure of the *Challenger.* By 1889, many important discoveries had been made in the Stazione through the study of marine larvae.

What Dohrn could see after Darwin was that, if not exactly as Haeckel thought, the development of embryos was in some measure related to that of species. It was at least in part in the light of his interest in marine biology, and in these minute creatures of the sea, that he could fathom the ultimate value of Darwin's theories.

Out of what intellectual depths of his own did he attain to such judgment? Haeckel may have stimulated his interest, but he thought his own thoughts for himself.

What had occurred during his earlier life as a scientist to sensitize him so to react to Haeckel's and Darwin's influence? We remember that he returned to concentration on zoology in 1862 after reconverting to it as a result of exposure to Darwin's ideas (see pp. 2–3 above). He took his doctor's degree in Breslau in 1865; his dissertation was devoted to Hemipteran anatomy. He was habilitated in Jena in 1868; his *Habilitationsschrift* was on a more general subject: The embryology and genealogy of Arthropods. Evolution, embryology, anatomy, taxonomy, and the interrelationships of all these were hence central in his thinking throughout his life. Although he had begun by concentrating on invertebrates, the evolutionary development of vertebrates from invertebrates became in due time, from 1875 on, the central focus in his thought; at the end of his life he devoted special interest to the related problems of the morphology and origins of the vertebrate head. His final scientific publication, a 293-page paper on the trochlear nerve, appeared in 1907,[46] two years before he died, 49 years after the publication of his first contribution to his father's journal.

Dohrn's strictly scientific contributions were always of high quality. His conclusion that the vertebrates evolved from annelids was of impor-

---

[44] J. Müller, "Bericht über einige neue Thierformen der Nordsee," *Arch. Anat. Physiol. wiss. Med.* [no vol. #] (1846): 101–110 [the discovery was made in autumn 1845].

[45] V. Hensen, "Ueber die Bestimmung des Planktons oder des im Meere treibenden Materials an Pflanzen und Thieren." *Komm. wiss. Meeresunters. Kiel,* 5. Ber. Jahrg. 12–16 (1887): 1–107.

[46] A. Dohrn, "Die Trochlearis." *Mitt. Zool. Stat. Neapel* 18 (1907): 143–436.

tance in its time, even though it later became superseded. His major personal contribution to evolution theory still stands as of crucial importance, having been completely absorbed into the body of argument on descent. That was his emphasis on *Funktionswechsel*, shift of function, as a factor in evolutionary change.

It was precisely during the time that Dohrn and von Baer were corresponding about Russian support for the Stazione that Dohrn was beginning to concentrate on vertebrate origins and writing on *Funktionswechsel*. His first mention of it to von Baer is found in his letter of 17.1.1875 (#27), when he sent on to him a draft of what would soon appear as a small monograph. He asked von Baer for his opinion of it, and for permission to dedicate it to him. The principle of *Funktionswechsel* states that an organ performing one function in an ancestral organism may subserve a different one in its descendants. We now accept this so completely that it no longer concerns us as dogma; we are so familiar, for instance, with the transformation of branchial and mandibular skeletal elements of lower vertebrates into the bones of the mammalian middle ear that we no longer feel constrained to defend the theory that accounts for the change.

On 8/20.2.1875 von Baer replied (#30), saying that he had agreed to the dedication in correspondence with Engelmann, the publisher. He admitted to having shaken his head at the idea that the annelids might have turned on their backs to become vertebrates, and he used an even stronger word (*anstössig* = shocked) in connection with his reaction to the idea that vertebrate appendages might be annelid respiratory organs transformed.

Von Baer then went on to comment on the fact that he was then writing for publication his own thoughts on Darwin's theory. "I cannot," he wrote in the letter to Dohrn, "refrain from believing that transformation is highly probable, but cannot accept Darwin's selection-hypothesis as a satisfactory explanation of it." Von Baer expanded further on this point in his letter to Dohrn of 11/23.6.1875 (#32); here he points out that he himself had suggested the likelihood of mutability in 1859. On 8 August 1860 he had written a letter to T. H. Huxley, in French, in which he said that "I have expressed the same ideas on the transformation of types or origin of species as Mr. Darwin. But it is only on zoological geography that I rely. You will find, in the last chapter of the treatise 'Ueber Papuas und Alfuren' that I speak of it very positively without knowing that Mr. Darwin was concerning himself with the subject." Even earlier, in 1834, he had admitted that transformation might occur, but in limited groups only: "If it is . . . permitted to imagine that antelope, sheep and goats . . . may have developed from a common original form, . . . yet I can on the other hand find no probability that all animals have developed from one another through transformation."[47]

---

[47] F. Darwin, ed., *Life and Letters of Charles Darwin, including an Autobiographical Chapter* (New York: D. Appleton, 1888), quotation from Vol. II, p. 329. [Von Baer wrote to Huxley in French and *Life and Letters* gives the extract from the letter in French.]; K. E. von Baer,

The general acceptance of Darwin's theory was not yet complete when the Stazione was founded. That is incidental; the role of natural selection in the process is again under dispute today.[48] Nonetheless, it was clearly as a result of the new emphasis on evolution that Dohrn was stimulated and impelled to found the Stazione. When work began there, the study of larvae in metamorphosing forms may well have contributed to the confirmation of evolution theory, but the truly pioneering work in its laboratories concentrated not on elucidating ancestries but rather on experimentation as a method of demonstrating and analyzing causal relationships of immediate, not ancient, import in controlling developmental events. At the turn of the century, these two modes of regarding causality in development seemed diametrically antithetical: Haeckel ridiculed the new experimentation. To him, the only causes of embryological events had to be phylogenetic. Now, a century or so later, evolution and embryonic development are both, as are most biological processes, open to analysis, and interrelated in molecular terms. While the accomplishments of early investigations at the Stazione may be viewed, in hindsight, as steps taken toward the science of the late twentieth century, their greatest significance was in the manner in which they contributed to the zoology, biology, and oceanography of their own day and von Baer as well as Dohrn contributed toward this end.

"Das allgemeinste Gesetz der Natur in aller Entwickelung. Ein Vortrag, gehalten . . . im Januar 1834" (?), [*sic*], in *Reden gehalten in wissenschaftlichen Versammlungen* (St. Petersburg: H. Schmitzdorff), I (1864): 37–74; quotation trans. from pp. 55–56.

[48] N. Eldredge, *Time Frames. The Evolution of Punctuated Equilibria.* (Princeton: Princeton University Press, 1989).

EDITORIAL REMARKS

Anton Dohrn corresponded with several outstanding scientists of his time, asking them for their support and approval for the foundation of a zoological station as an independent research institute for studies in marine biology. Letters to Anton Dohrn and documents relating to the foundation, management, and scientific activity of the Zoological Station have survived two world wars with only minor losses. The Archives of the Zoological Station were formally established in 1969; its rich holdings have since then been open to scholarship.

As part of an extensive editing program, Dohrn's letters to and from Charles Darwin (Groeben, 1982), Emil du Bois-Reymond (Groeben, 1985a), and Rudolf Virchow (Groeben and Wenig, 1992) have already been published. His correspondence with one more of the founders of modern biology seems to be an appropriate contribution to the bicentennial of the birth of Karl Ernst von Baer.

Of the 41 letters and documents published here, 12 are held in the private archives of the Dohrn family at Colli sul Velino (Rieti), Italy [FA], one at the Bayerische Staatsbibliothek München, Germany [FAM], and 28 are held today at the Universitätsbibliothek Giessen, Abteilung Handschriften, Nachlass von Baer [UBG] where the letters of Anton, Carl August, and Heinrich Dohrn are part of a collection of about 4,000 to 5,000 letters, which include both letters to von Baer from ca. 900 correspondents and from von Baer (mostly copies) to about 108 addressees. All documents held in von Baer's papers are stamped "Bibliothek der Ludwigs-Universität Giessen. Buchstempel." About the history of the von Baer Papers, see Introduction, p. 1.

Von Baer's letters to Anton Dohrn have always been held by the Dohrn family. For the respective provenance of each letter see the chronological list on pp. 29–30. All the documents are published with permission.

Twenty-two letters are from Anton Dohrn, three of them with additions by Carl August Dohrn, Anton's father; four letters to von Baer are from C. A. Dohrn, one from Heinrich Dohrn, Anton's brother, while thirteen letters and one telegram are addressed by von Baer to Anton Dohrn. As has been mentioned in the Introduction (pp. 9–10), the letters from Carl August and Heinrich Dohrn have been included in this collection because they document the background and beginning of von Baer's relation to the Dohrn family long before Anton and his Neapolitan enterprise entered the stage.

All the letters are manuscripts, but only the Dohrn letters are autographs. With the exception of one (#7 below) von Baer's letters are all signed by him but have been dictated to various persons. Four different amanuenses can be distinguished: letters 9, 15, 17, 18, 23, 24, and 26 are by one hand, letters 11 and 21 by another, letters 30 and 37 are written by a third person, and letters 32 and 35 by a fourth.

In 1895 Anton Dohrn started to write the history of the Naples Station (see note 20 to letters); for that purpose he selected relevant documents from his rich archives which he then listed and summarized chronologically according to years, or as they related to famous scientists. For von Baer, Dohrn selected eight letters (1873–1875), numbered by letters (a–h); two more are listed in Dohrn's selection for 1874 and two for 1875. One letter (#7, see note 30 to letters) has survived only in a copy by Anton Dohrn, while the telegram from von Baer (letter 21) has not been preserved by Dohrn (the original is located at Giessen).

With the exception of letter 3 (see note 11 to letters), all the following letters are unpublished, although parts of von Baer's letters have frequently been quoted, in particular by Heuss for his biography of Anton Dohrn (Heuss, 1962). It was therefore desirable to present the letters in their original German version. This meant that two dangers had to be avoided. Presenting only the German texts, with English notes and introduction, would have created a hybrid book that would exclude potential readers fluent in only one of the two languages; another solution, namely giving both the original and the translation face to face, or one version in smaller print following the other, would have encouraged skipping on the part of readers. It was therefore decided to present a unified English edition of the annotated letters and the introduction, and to publish the original German texts, without explanatory notes, in an appendix.

The letters are arranged chronologically. The letter numbers of both the English and the German versions are followed by the same essential information, namely author, addressee, date, and place. These data are given in square brackets when they had to be deduced from the contents of the letters or from other sources. The chronological list on pp. 29–30 will facilitate the location of the English and/or German version of the letters.

The original German letters have been carefully transcribed with regard to spelling, grammar, and punctuation. Notes to them are marked by letters and refer only to text critical observations. Anton Dohrn's letters to von Baer have unfortunately been pasted together on their left border, parts (words or letters) of the text are therefore lost at times; obvious and possible reconstructions as well as missing parts are marked by square brackets [ ], [?], [. . .]. Misspellings are followed by [*sic*]. The sign for the German monetary unit "Thaler," often used by Dohrn, is spelled out as [Taler] or, in the German letters, as [Thaler].

At the period of the correspondence von Baer's first name "Karl" was also spelled "Carl"; von Baer himself always signs his letters as "Dr. K. E. v(on) Baer" while Anton Dohrn generally uses "Carl" (see letters 10, 20, and note 80). Russian names are given in their modern English transcription, except in the German letters. Printed letterheads are shown in italics.

Unless otherwise indicated, cross-references to notes refer to notes to the correspondence. Translations in the notes are my own, as edited by Jane Oppenheimer in some cases.

Acknowledgments

Thanks are due to Antonietta and Peter Dohrn, and to the Universitätsbibliothek Giessen for permission to publish von Baer's and the Dohrns' letters respectively. I am grateful to Bernd Bader (Giessen) for patiently checking my photocopies against the original letters of Anton, Carl August, and Heinrich Dohrn and for his suggestions in reconstructing the missing parts, and to Ilse Jahn (Berlin), Erika Krausse (Jena), and Dorothea Kuhn (Marbach a.N.) for helping with bibliographical information. Photographic prints for the illustrations were provided by the Photo Laboratory of the Stazione Zoologica "Anton Dohrn."

I am grateful to the Fritz-Thyssen-Stiftung for generously supporting through the Max-Planck-Gesellschaft the initial stages of this project.

Dr. Oppenheimer received much technical help, for which she is grateful, in the preparation of her Introduction and its notes, from Anne Pringle, Science Librarian at the Bryn Mawr College Library, and from Roy Goodman, Curator of Printed Materials of the Library of the American Philosophical Society in Philadelphia.

CHRISTIANE GROEBEN
Stazione Zoologica "Anton Dohrn"
Villa Comunale
80 121 Naples, Italy

LIST OF ABBREVIATIONS

AD	Anton Dohrn
ASZN	Archives Stazione Zoologica di Napoli
CAD	Carl August Dohrn
FA	Dohrn Family Archives, Colli sul Velino (Rieti)
FAM	Dohrn Family Archives, Bayerische Staatsbibliothek München
HD	Heinrich Dohrn
UBG	Universitätsbibliothek, Abteilung Handschriften, Justus-Liebig-Universität Giessen
vB	Karl Ernst von Baer

CHRONOLOGICAL LIST OF THE LETTERS

				PAGE	
[1] CAD to vB	10.9.1862	Stettin	UBG:1	31	*97*[1]
[2] HD to vB	3.6.1863	Stettin	UBG:2	32	*98*
[3] CAD to vB	6.11.1864/ 12.2.1865	Stettin	UBG:3	34	*99*
[4] AD to vB	30.12.1869	Stettin	UBG:4	35	*100*
[5] CAD to vB	2.10.1871	Stettin	UBG:5	38	*102*
[6] AD/CAD to vB	2.1.1873	Stettin	UBG:6	40	*103*
[7] vB to AD	15./27.1.1873	Dorpat	ASZN:*Ha*	42	*105*
[8] AD to vB	8.2.1873	Naples	UBG:7	44	*106*
[9] vB to AD	18.8.1873	Dorpat	FA:*Ba 1034* (a)	47	*109*
[10] AD to vB	7.9.1873	Brighton	UBG:8	48	*110*
[11] vB to AD	31.8./12.9.1873	Dorpat	FA:*Ba 1033* (c)	50	*111*
[12] CAD to vB	18.9.1873	Stettin	UBG:9	51	*113*
[13] AD to vB	19.9.1873	London	UBG:10	52	*113*
[14] AD to vB	29.9.1873	Stettin	UBG:11	54	*114*
[15] vB to AD	18.10.1873	Dorpat	FA:*Ba 650* (b)	55	*115*
[16] AD to vB	23.1.1874	Naples	UBG:12	56	*116*
[17] vB to AD	24.1./5.2.1874	Dorpat	FA:*Ba 166* (#8)	58	*117*
[18] vB to AD	23.3./5.4.1874	Dorpat	FA:*Ba 201* (#27)	60	*119*
[19] AD to vB	16.4.1874	Naples	UBG:13	61	*119*
[20] AD to vB	28.5./[9.6.1874]	Petersburg	UBG:14 [Telegr.]	64	*122*
[21] vB to AD	28.5./[9.6.1874]	Dorpat	UBG:15 [Telegr.]	64	*122*
[22] AD to vB	11.6.1874	Petersburg	UBG:16	65	*123*
[23] vB to AD	25.7./6.8.1874	Dorpat	FA:*Ba 1035* (d)	67	*125*
[24] vB to AD	26.7./7.8.1874	Dorpat	FA:*Ba 1036* (e)	69	*126*
[25] AD/CAD to vB	10.8.1874	Stettin	UBG:17	70	*127*
[26] vB to AD	[10.1.1875]	[Dorpat]	FA:*Ba 1037* (f)	71	*128*
[27] AD to vB	17.1.1875	Naples	UBG:18	73	*129*
[28] AD to vB	6.2.1875	[Naples]	UBG:19	75	*131*
[29] AD to vB	20.2.1875	Naples	UBG:20	76	*132*
[30] vB to AD	8./20.2.1875	Dorpat	FA:*Ba 1038* (g)	76	*132*

[1] Page numbers shown in this column refer to the German original of the letters.

					PAGE	
[31]	AD to vB	1.4.1875	Naples	UBG:21	77	*133*
[32]	vB to AD	11./23.6.1875	Dorpat	FA:*Ba 1039* (h)	78	*134*
[33]	AD to vB	8.7.1875	Hökendorf	UBG:22	81	*137*
[34]	AD to vB	18.8.1875	Stettin	UBG:23	83	*139*
[35]	vB to AD	16./28.8.1875	Dorpat	FAM:1 (#62)	86	*142*
[36]	AD/CAD to vB	3./4.9.1875	Stettin	UBG:24	87	*143*
[37]	vB to AD	17.9.1875	Dorpat	FA:*Ba 296* (#68)	89	*145*
[38]	AD to vB	15.10.1875	Berlin	UBG:25	90	*145*
[39]	AD to vB	15.10.1875	Berlin	UBG:26	91	*147*
[40]	AD to vB	22.10.1875	Petersburg	UBG:27	92	*147*
[41]	AD to vB	2.5.1876	Naples	UBG:28	93	*148*

KARL ERNST VON BAER – ANTON DOHRN CORRESPONDENCE

[1] Carl August Dohrn[2] to von Baer 10 September 1862, Stettin

Academician von Baer
Excellence

Most honored friend,

On my return in June from a trip to England and France, I found your esteemed letter regarding the sea water to be sampled on the east coast of Rügen.[3] I have commissioned several friends to do so, since several obligations prevented me from going to Rügen myself. But the truth of the saying *aide Toi* proved itself again, the persons I asked returned *without* the desired seawater and the only thing I could do was to take advantage of the visit of a naturalist friend from Paris, to go with him at the beginning of September on an entomological hunt to Stubbenkammer, and on this occasion also to fill the two enclosed bottles.

Although the samples have not been taken, as you had wished, near Arcona, but nevertheless not far from there, and since no fresh water (with the exception of highly insignificant brooks) flows into the Baltic Sea near Stubbenkammer, the salt content will probably be almost identical with that of the seawater close to Arcona.

In a few days I will go with two of my sons, among them Heinrich who

---

[2] Carl August Dohrn (CAD; 1806–1892; see Introduction pp. 6–7), "world-traveler, singer, entomologist" (K. Dohrn, 1983, p. 30). Through exchange and acquisition, his entomological collection grew to about 40,000 specimens, a remarkable achievement for a private person. He edited the *Stettiner Entomologische Zeitung* (1840–1887) and *Linnaea entomologica* (1852–1866) and was president (1843–1887) of the Entomological Society of Stettin, founded in 1837; in 1862 the University of Königsberg conferred on him a PhD *honoris causa*. CAD was also an enthusiastic and gifted correspondent; his letters are saturated with quotations, allusions and puns. His love for Goethe, whom he visited on 17 September 1831 in Weimar, was passed on to his youngest son Anton (H. Dohrn, 1892; Heuss, 1962, pp. 23–49; K. Dohrn, 1983, pp. 30–81).

[3] Von Baer must have written to CAD before his departure to the Sea of Azov (May 1862). Stimulated by similar experiments in France, von Baer collected information on the presence and breeding of oysters in the past, and of the salt-percentage in several parts of the Baltic Sea [K. E. von Baer, "Ueber ein neues Projekt, Austern-Bänke an der Russischen Ostsee-Küste anzulegen und über den Salz-Gehalt der Ostsee in verschiedenen Gegenden (mit einer Karte, welche den Salz-Gehalt der Ostsee in verschiedenen Gegenden anzeigt). Und Nachträge," *Bull. Acad.* 4, (1862): col. 17–47; 119–149]. The results from the samples from the Island of Rügen, sent to him by CAD, seem not to be mentioned anywhere in print. Nor is anything known about Heinrich Dohrn's report on the fauna of the same region (see letter 2).

has the honor of being known to you, to the meeting of German naturalists at Carlsbad.[4] There is a possibility that we will meet you there, which would be a pleasure for us. Because of this I address this letter to the Imperial Embassy so that the bottles will not undertake the voyage in vain.

They will leave here next Saturday (13 September new style)[5] on the steamer *Trave*, Captain *Heidtmann*, for Petersburg and the Imperial Academy should therefore be kind enough to have the bottles which are in a wooden box addressed to the *Academy of Sciences*, picked up on the arrival of the Trave (16 September new style) at its landing place in Petersburg.

Heinrich will try his best to comply with your wish to obtain information on zoological products found on the east coast of Rügen. Such things present strange difficulties because most of the time one receives insufficient information from nonspecialists and one has to ask them specifically almost everything, which allows all sorts of errors to slip in.

Commending myself to your kind remembrance and with the request that you convey my respects to your wife,[6] I am,

Sincerely yours,
Dr. C. A. Dohrn

Stettin, 10 September 1862.

---

[2] Heinrich Dohrn[7] to von Baer
3 June 1863, Stettin

Stettin, 3 June 1863

Highly honored Sir, and Benefactor,

Already during the winter I should have reported to you on what I have accomplished up to now in the matter of the fauna of the Baltic Sea, but since I wanted to send in my results personally I have hesitated up

---

[4] The 37th annual meeting of German Naturalists and Physicians took place at Carlsbad in September 1862. Heinrich Dohrn (see note 7) had visited K. E. von Baer in St. Petersburg in 1861. The second son, mentioned by CAD, must have been Anton Dohrn, who at that time had just finished his first semester at Jena (summer 1862), while von Baer had just returned from a preliminary expedition to the Sea of Azov (May-September 1862).

[5] New style = Gregorian calendar; old style = Julian calendar, the latter in use in Russia until the Revolution. Until 1800 the Julian calendar was 11 days behind the Gregorian, after 1800 the difference increased to 12 days.

[6] Auguste von Baer, née von Medem (1799-1864).

[7] Heinrich Dohrn (1838-1913) had studied natural science in Bonn, Zurich and Berlin (PhD, Berlin, 1861). He never had to earn his living. During his numerous travels, he increased his father's entomological collection, while he himself specialized on conchylia (sea shells). With strong political ambitions, he served the City Council of Stettin (1869) and was elected member of the Prussian Parliament (1873-1879) and the Reichstag. A generous patron, he gave the Pomeranian Museum to Stettin. In 1887 Heinrich succeeded his father as president of the Entomological Society and as administrator of the family property at Hökendorf near Stettin (K. Dohrn, 1983, pp. 84-95).

to now to do so. Moreover, I have also had an eye-disease, which was insignificant in itself but had however the well-known persistence of diseases of this organ, so that for weeks I had to disregard all activities, primarily reading and writing, that would have been trying for the eye. Now, thank God, I am fine again, and after having dealt with an entomological matter I shall first of all try to finish my part of the Baltic sea fauna; I am only afraid that it will remain extremely fragmentary; at least for the lower animals, with the exception of the mollusca, the material is more than scanty.

I have to answer one more question that you have put to me—regarding my collection of conchylia. The latter indeed begins to become one of the big private collections since it includes about 10,000 species; I shall therefore probably be able to supply you as far as possible with information about the small items from the Sea of Azov,[8] but I want to add right away that the fauna there is very little known and will therefore contain much that has not yet been described. The Greek archipelago and the west coast of Asia Minor are almost *terra incognita* for conchylia.

If you would therefore entrust to me for some weeks for examination the items you brought along, it will be a pleasure for me to work through the fauna there according to your wishes.

Best greetings from my father and he is counting very much on seeing you here in September for the meeting of naturalists[9] and to grace our meeting in a special way; if you do not come he would make you the subject of a satirical poem; I on my part hope that you are not going to expose yourself to such an eventuality, and can only say that I would be extremely glad to see you here.

Your obedient servant
Heinrich Dohrn

---

[8] In order to investigate the causes of the increasing silting of the Sea of Azov, the Russian Academy of Sciences, together with the Geographical Society, planned an expedition under the direction of N. J. Danilevskij (1822–1885). Two years were planned for the hydrographical, geological, botanical and zoological investigation. K. E. von Baer had agreed to direct a preliminary expedition. From June to September 1862 he stayed at the Azov Sea where he also assembled a considerable collection of invertebrates (von Baer, 1866, pp. 437–439; Raikov, 1968, pp. 256–259). In a letter that has not been preserved, von Baer seems to have asked Heinrich Dohrn, known to him as an expert on conchylia, to classify some of his findings. Nothing is known about H. Dohrn's participation in this venture.

[9] Thanks to the initiative of C. A. Dohrn as president of the Stettin Entomological Society, the 38th meeting of the German Naturalists and Physicians was held at Stettin from September 18 to 24, 1863; university towns were usually chosen for those meetings. The Stettin meeting earned a special historical significance through the fact that on 19 September 1863 Ernst Haeckel gave a lecture entitled "Über die Entwickelungstheorie Darwins" (C. A. Dohrn, 1864, pp. 17–30) advocating Darwin's theories in public in Germany for the first time.

[3] Carl August Dohrn to von Baer
6 November 1864/12 February 1865, Stettin

To the Academician
Dr. Jubilatus von Baer
Honorary Member of the Entomological Society of Stettin.[10]

Excellency!

The Pomeranians always have to have something out of the ordinary! Where others prick with bayonets, they beat with clubs, and where others stick anxiously to day and hour on 9 September, they congratulate on the Doctor's jubilee, in all leisure, only on 6 November.[11] *Honni [sic] soit, qui mal y pense!* But then why did the Giessen meeting of the naturalists have to fall just in September and why is the official report of the Stettin meeting complete only now?[12]

In respectfully presenting one copy of the latter to our celebrated honorary member we attach to it the sincere wish that the *ursa major*[13] might still shine for a long time over our society in physical comfort and with the undiminished intellectual vigor that distinguishes our hero in science from many others. On this a foaming glass has been emptied today with all our heart during the celebration of our commemoration day!

Stettin, 6 November 1864.

In the name and on behalf of the
Stettin Entomological Society
The President
Dr. C. A. Dohrn

Dr. Heinrich Dohrn, who is setting out tomorrow on a scientific trip to the Cape Verde Islands, joins his cordial sincere greetings to mine.

Vertas quaeso.

---

[10] Von Baer published only a few minor, mostly trivial papers on insects; far from trivial, however, he also described and defined paedogenesis in insects. In May 1860 he was involved in the formation of the Russian Entomological Society. At its opening he gave one of his most important philosophical talks, "Welche Auffassung der Natur ist die richtige, und wie ist die Auffassung auf die Entomologie anzuwenden?" *Horae Soc. Entomol. Rossicae (Petropoli)* 1 (1861): 1–45; reprinted in von Baer, 1864–1876, vol. 1, pp. 237–284.

[11] On 29 Aug./9 Sep. 1864 the 50th anniversary of von Baer's doctorate had been celebrated in St. Petersburg. CAD's letter was included in the publication of the speeches, messages, and decorations with the introductory remark: "Significantly *post festum*, but still not too late, the following letter arrived from Stettin" (translated from the German; [von Baer], 1865, pp. 127–128).

[12] C. A. Dohrn, 1864.

[13] Referring, obviously, to the constellation, but also to von Baer: the German meaning of Baer = Latin, ursus (= English, bear), and a noble (von) bear being then a *ursus major.*

Stettin, 12 February 1865

That the preceding tardiness has fallen again in the abyss of procrastination is less *my* fault than that of the blockading Danes[14] and the disappointed expectation that someone would bring this letter and the appertaining promemoria overland into your honored hands, but then, however, the person did not pick it up. But according to the wise words *Better late than never* I send you today at least this letter and when I shall happily have made the forthcoming trip to Paris and Palermo and shall have disposed myself back *ad penates* in May *Diis faventibus,* the D.[15] will not again lay obstructive eggs into the steam navigation path and make trouble by waylaying again the copy of the naturalists' report destined for you.

In any case I can tell you today that Dr. Heinrich has *feliciter* arrived on the Cape Verde Islands and that after a sojourn of some four weeks has already released an eight page letter according to which he is quite satisfied with his adventures and results, although more with the malacozoological part than with the entomological; but that was to be expected *a priori.* The tropical heat suits him pretty well and he achieves considerable accomplishments in doing away with oranges.

Treasured friend, keep him and me in your kind favor and accept my hearty greetings.

Yours truly,
Dr. C. A. Dohrn

[4] Anton Dohrn to von Baer
30 December 1869, Stettin

Stettin, 30 December 1869

Dear Sir!

Fortunately I am not completely unknown to you—which spares me the necessity of first presenting my credentials.

I am impelled to write this letter by a plan that has already been made known to you by my friend Micloucho-Maclay.[16] I intend to found zoo-

[14] During the German-Danish war of 1864 for Schleswig-Holstein which terminated with the signing of the peace-treaty of Vienna (30 October 1864), Danish warships had blocked off the German sea-ports.

[15] The Devil.

[16] Together with his Russian friend, Nikolai N. Micloucho-Maclay (1846–1888), Anton Dohrn had spent the winter of 1868–1869 at Messina for studies in marine biology (see Introduction, p. 3). Before that Dohrn had worked in Helgoland (August 1865 with Haeckel) and at Millport, Scotland (1867, 1868) as guest of the Scottish amateur zoologist David Robertson (1806–1896), whereas Maclay had already gained seaside experience on an excursion to the Canary Islands (1866–1867) with his teacher Ernst Haeckel. The two friends decided to cover the globe with a system of zoological stations where traveling scientists

FIGURE 1. Anton Dohrn at the age of 23, photograph by H. Graf, Berlin, 1863 (*ASZN*).

logical stations which offer to zoologists arriving at the seaside all hand tools, a boat, drag net, ropes, as well as aquaria and lastly even a house with living room and study for which the user will have to pay a modest sum.

I shall return to Messina next fall in order to occupy myself in detail with the embryology of the annelids. Before then I want to collect enough money to be able at the same time to undertake the building of the first station. I estimate the costs at the moment at 2,000 Taler which I plan to raise through collections.

I need not explain to you, dear Sir, what help it would mean for me if you would express in a letter to me your participation in and approval of the enterprise. Your name would legitimize me everywhere and I could hope thereby to find a favorable consideration with persons who do not know me and could therefore not judge whether I am planning something useful.

Since through my studies on crabs and other marine animals I could gather sufficient experience on what must necessarily be acquired and installed for a zoological laboratory, I believe that you might boldly dare to express a vote of confidence on my behalf. Through many connections with colleagues I hope later to succeed in establishing in the station a zoological library which is of such great importance for the traveling zoologist.

I am not going to tire you with lists of all that could be done: you know that better than I do, and it well becomes me to follow your kind advice instead of making proposals.

May I therefore once again ask you for this advice? You would not give it to a person who belongs into Goethe's category of "Schiefohren"[17]– and I would be thoroughly grateful.

I am also sending you the first part of my work[18] that presents itself to you without any other claim than to introduce the author to its patriarch as active in the paths of embryology. I hope that two or three more will follow during the summer.

---

would find top working facilities. Dohrn rented two rooms in Messina where he left his portable aquarium and microscope, his books and diaries. The "Zoological Station of Messina" existed for one year until in January 1870 Dohrn decided to transfer the Station to Naples (see note 20).

Maclay traveled from Messina to the Red Sea where he studied the local sponge fauna before returning to Moscow and Petersburg in the summer of 1869. At the meeting of Naturalists in Moscow his report on the Messina Station "caused a sensation" (letter from Maclay to Dohrn, 16 Sept. 1869, I. Müller, 1980, pp. 54–55). In St. Petersburg he worked at the Academy on a collection of Asiatic sponges and on this occasion he probably reported to von Baer on their enterprise. Maclay never tired of propagating the idea of zoological stations. In 1878 he founded a zoological station at Watson's Bay, near Sydney, Australia, before turning his scientific interest to anthropology, in particular to the study and protection of the Papuans in New Guinea (I. Müller, 1980).

[17] Literally: crooked ear (tin ear); for Goethe a person without judgment or sensitivity.

[18] A. Dohrn, *Untersuchungen über Bau und Entwicklung der Arthropoden*. 1. Heft (I–V). Leipzig: Engelmann, 1870. 103 pp., 9 pl. Delivery already began in November 1869. The copy in the library of the Naples Station is signed: "Anton Dohrn 1869."

Please forgive this predatory attack on your time by a young man who deeply links the pleasure of his profession with his admiration for your person.

Yours faithfully,
Anton Dohrn
youngest offspring of the family.

My father sends his heartiest greetings, so does my brother Heinrich! From 8 January, I shall again be active as a lecturer in Jena.[19]

---

[5] Carl August Dohrn to von Baer
2 October 1871, Stettin

Stettin, 2 October 1871

Dear friend and benefactor,

On behalf of my youngest son Dr. Anton, up to now *Privatdozent* [lecturer] at the University of Jena, I have the honor to submit to you the enclosed project of his new enterprise.[20] If it turns to water then it will at least be a comfort to know that it is salt water where perhaps inadvertently some Attic crystals will also slip in. Since Anton has already sought and found opportunities to discuss this matter with zoological

---

[19] During the winter semester of 1869–1870 Dohrn gave a two-hour course on "Entwickelungsgeschichte der Ringelwürmer und Gliedertiere" with 12 students registered, and sometimes as many as 6 guests (Uschmann, 1959, p. 86).

Dohrn had received his PhD in November 1865 at Breslau with Eduard Grube (1812–1880); his dissertation "Zur Anatomie der Hemipteren" was published in the *Stettiner Entomologische Zeitung* 24 (1866): 321–352. He was habilitated in the spring of 1868; his habilitation-presentation was published as a monograph: "Studien zur Embryologie und Genealogie der Arthropoden" (Leipzig, 1868). In the summer semester of 1868 he started to lecture at Jena University.

[20] A few days after writing his letter to von Baer (30 December 1869, letter 4) Dohrn visited the new public aquarium in Berlin. The entrance fees to a similar aquarium in connection with his Zoological Station would give him, Dohrn reasoned, the necessary means for paying a permanent assistant and cover the running expenses of the Station. His final choice therefore fell on Naples, where more tourists and potential visitors could be expected.

In his letter CAD enclosed the small pamphlet: [A. Dohrn], *Hauptgrundzüge des Contractes zwischen Dr. Anton Dohrn und der Stadt Neapel betreffend die Errichtung der Zoologischen Station.* Leipzig: Breitkopf und Härtel, [1871], 8 pp. The publication also contained information on the sea-water system, the laboratories and collections, the library and ships, staff and publications, all of which materialized much as Dohrn had planned long before actual construction began (March 1872). The contract with the City of Naples underwent several changes and delays before it was signed in December 1875, long after the Institute had opened in September 1873.

According to Dohrn (A. Dohrn, [Memoirs, History of the Naples Station (1868–1875), 1895–1909], fragment, 147 pp., ms. / typescript; typescript p. 59) his father had also enclosed a collection of letters from "authorities in science," published by Dohrn in September 1871 (A. Dohrn, [1871a]). It contained letters from Emil du Bois-Reymond, Carl Gegenbaur, Ernst Haeckel, Hermann Helmholtz, Rudolf Leuckart, and Carl Vogt.

celebrities (Huxley, Darwin, van Beneden, Leuckart, Dubois[21] etc. etc.), since they have jointly and separately expressed themselves as highly satisfied with the idea and modalities, I dare hope that you, my most esteemed friend, will also not deny your benevolent interest in this apparently venturesome project. The venturesomeness consists less in the idea than in the real difficulty of convincing Neapolitans of the fact that one is willing, for a purely scientific purpose, to risk effort and "ready money!!!" The Mayor and councillors have not been ashamed to say bluntly that the so-called aquarium is probably only meant to disguise a hotel if not a brothel.

Anton has recently been permanently on the road, he visited the new (less than perfect) installations at Brighton[22] (where they also have prepared a tank for a whalefish [sic]) and the meetings of scientists in England, Scotland, Rostock[23]; he has prepared contracts with suppliers of machines, plate glass etc. and is now on his way to the archaeological meeting in Bologna[24] before embarking on the matter in Naples.

---

[21] Thomas Henry Huxley (1825-1895), British zoologist and comparative anatomist, friend and public supporter of Charles Darwin. From its very beginning Huxley generously supported Dohrn with help and advice in creating the Zoological Station. A warm friendship linked Dohrn to the whole Huxley family; their rich correspondence illustrates their common scientific and literary interests. Dohrn also owed his introduction and visit to Darwin on 26 September 1870 to Huxley (Groeben, 1982, pp. 93-94).

Charles Darwin (1809-1882) appreciated Dohrn's scientific work. He also considered the creation of a zoological station as a great service to science; his support included gifts of books and money (Groeben, 1982).

Dohrn had met Pierre Joseph van Beneden (1809-1894), professor of zoology at Louvain (1836) and member of the Belgian Academy of Sciences, at the annual meeting of the British Association in Liverpool (September 1870); Dohrn asked him for moral support for his enterprise in Belgian scientific circles (A. Dohrn, [Memoirs, see note 20], ms. pp. 37b-38b).

Rudolf Leuckart (1822-1898), professor of zoology at Leipzig since 1869, supported the Station, which he visited in 1884, from the very beginning (A. Dohrn, [1871]).

The German physiologist Emil du Bois-Reymond (1818-1896) generously used his considerable influence as a member of the Berlin Academy of Sciences in favor of the Zoological Station (Groeben, 1985a).

[22] There is no evidence of Dohrn's visit to Brighton in 1871. Construction of the local aquarium there began in the fall of 1869; the aquarium was opened in August 1872 for the annual meeting of the British Association, which Dohrn also attended (W. Saville Kent, "The Brighton Aquarium," *Nature* 8, 1873: 531-533). One of the inconveniences of the water supply system used in Brighton was that the tanks were not connected to each other, thus excluding any circulation of water. A well-functioning system of permanent water exchange, resulting from an interconnecting pipe system for all the tanks, was the invention of the British technician and engineer, William Alford Lloyd (1815-1880); he used it first in Hamburg (1868), and then in London (Crystal Palace, 1871) and Naples (1872). While the London aquarium in the end had cost £12,000, building costs for Brighton amounted to £100,000 (Lloyd to Dohrn, 16.2.1872, *ASZN: Ha 1872.69a*).

[23] In September 1870 Dohrn had attended the annual meeting of the British Association at Liverpool; this was of particular importance for Dohrn insofar as the General Assembly unanimously voted the nomination of a committee "for the purpose of promoting the foundation of Zoological Stations in different parts of the world, recognizing the foundation of a Zoological Station at Naples as a decided step in this direction" (*Rep. Brit. Ass. Adv. Sci. Liverpool 1870*, 1871, LXI). The following year (1871) Dohrn attended in August the annual meeting of the British Association in Edinburgh and in September the 44th meeting of German Naturalists and Physicians at Rostock.

[24] In October 1871 Dohrn moved definitely to Italy. On his way to Naples he attended

Since I believe that the last war, as a safety valve, has permitted the escape of a large quantity of reeking gases for several years, I nourish the hope that the financial calculations are not construed in too sanguine a way and that when a few quiet years will have set the Zoological Station going, it will be able then to speak for itself.

My oldest son, Dr. Heinrich, who had the honor to present himself to you in Petersburg and who since 1863 directs and improves *pro virili* the zoological materials in the *Pommersches Museum*, founded at that time, is expected back today or tomorrow from an excursion to the United States where Agassiz[25] and Dr. H. Hagen[26] have received him in a most friendly way and have so generously overwhelmed him with donations for the said museum that he could not place everything.

Thus—as you can see—the Dohrn family documents an unexpected quantity of beastly things—*eheu*!

Personally I am fine, thank God, and I am hoping sincerely that you and your family are equally well. Perhaps you could delight with a few friendly lines

Your most sincere admirer,
Dr. C. A. Dohrn

[6] Dohrn/CAD to von Baer
2 January 1873, Stettin

Stettin, 2 January 1873

While I was reading yesterday the passage in your autobiography,[27] Excellency, where you give an account of the beginning of your embryological studies under Döllinger,[28] I thought that it should appear to

the 5th International Congress of Prehistorians in Bologna. Rudolf Virchow (1821-1902) had been invited there as the only official German representative. On this occasion Dohrn came to know his former teacher much better than before on a personal level. Later Virchow repeatedly used his considerable political and scientific influence in favor of the Zoological Station (Groeben and Wenig, 1992). At Bologna, Dohrn could also establish first contacts with Italian scientists such as Salvatore Trinchese (1836-1897), who in 1880 succeeded Paolo Panceri (1833-1877) as professor of comparative anatomy at the University of Naples.

[25] Louis Agassiz (1807-1873), Swiss zoologist, geologist, and oceanographer, since 1846 in the United States, founder of the Harvard Museum of Comparative Zoology at Cambridge, Mass. (see note 53).

[26] Hermann August Hagen (1817-1893), professor of entomology at Königsberg, went to Cambridge, Mass. in 1867 as assistant of Louis Agassiz; in 1870 he was nominated professor of entomology at Harvard College. A lively and rich correspondence between Hagen and CAD started in 1844; their scientific and personal relations were later extended to include both Heinrich and Anton Dohrn.

[27] Von Baer, 1866.

[28] In 1815, von Baer arrived in Würzburg where he acquired his first practical experience in comparative anatomy, studying with anatomist Ignaz Döllinger (1770-1841) (von Baer, 1866, pp. 165-169).

you to be a continuation of your own intentions and projects when at present in Naples the finishing touches are given to the completion of a big building that has virtually been born from the need to pave the way and further the means for embryological research. Excellency, you know how scarce our knowledge is of the development of the fishes, that the embryology of cephalopods, worms, echinoderms—in short, of almost all marine animals, has still to be made, even though the last decades have done much more in this very direction than have previous times.

I believe that the biggest difficulties that these important investigations have gone through for many different reasons have been if not eliminated, at least much diminished, by the construction and equipment of the "Zoological Station," and at the same time I am trying to organize the Zoological Station in such a way as to grant as many zoologists as possible the opportunity of its use at the lowest possible expense. I am about to make arrangements with the German governing bodies, which are supposed to grant on one hand available means to the Zoological Station, on the other great facilitation to German zoologists for their research in Naples,[29] and once this has been accomplished for Germany, I shall try to achieve it for other countries, and in particular also for Russia, that is sending so many and such able men specifically to Naples.

However, I would have preferred to announce all this to your Excellency as already definite,—but your Excellency will permit me to address a second letter to you in hopes of making up for this. But since I visited my parents at the end of last year and my father asked me whether I had already informed you of the progress of my enterprise, this question coincided with my own inner desire to let you have at least a short report that the whole matter is approaching its completion.

Excellency, you will not interpret this as an importunity of a young expert, but recognize it as the wish to tell the old master how much today's generation considers itself as his descendants in science, and although it will be difficult if not impossible to achieve again such great and pioneering work, we shall nevertheless continue to walk *viribus unitis* on the path you have prepared and to strive for the goal that you have set out for us.

---

[29] Dohrn refers to what has been called the "table-system." Already during construction Dohrn realized that the entrance fees to the public aquarium would never cover the total running expenses of the Station. He first considered a joint-stock venture, but arrived instead at a solution that would guarantee a certain financial security without limiting his independent power of decision. Dohrn decided to rent laboratory space or "worktables" to government and scientific institutions (see Introduction, especially p. 15). For an annual fee the contract partner acquired the right to send one scientist for one year to Naples where he would have free access to laboratories, stock room supplies, library, living material etc. Thanks to Dohrn's diplomatic skills this new way of scientific organization, namely offering top research facilities and complete freedom of decision in their use by investigators, worked perfectly. It has been adopted by several other institutions such as the Marine Biological Laboratory at Woods Hole, the Rockefeller Institute for Medical Research, New York, and the Kaiser Wilhelm Institutes in Berlin, although each of them employed different methods of gathering funds.

Excellency, please take this letter as the expression of my desire that I might be granted to give you often New Year's reports on the welfare and the activity of the Zoological Station; this is mine and the Zoological Station's heartiest wish.

My father sends his most cordial greetings and my brother Heinrich also sends his regards.

With due respect and best wishes for the New Year
Yours truly,
Anton Dohrn

[CAD:]
With heartiest greetings and wishes from yours respectfully, who last year in Florence for unknown and uninvestigated reasons almost bit the dust, but who thanks be to God already kicks out again *more solito* before and behind, skates and practices similar *juvenilia* with or without gracefulness. The noble passion for the lousy vermin flourishes as ever, consignments fly to and from Sydney, New Guinea, Hottentot-land, Cambridge-Massachusetts, in short the sixlegged freemasonry is in full bloom.

It is only natural that I am interested in Anton's enterprise in every regard, either with my purse or with good scientific hope. Dr. Heinrich, who has had the honor to glory in your personal benevolence, also continues his zoological studies as director of the beastly department of the Pomeranian Museum, although as city councillor he is at the moment preoccupied with new harbor and dock facilities.

Repeating with all my soul the hope that the New Year might preserve for you untroubled serenity and health,

Yours sincerely,
C. A. Dohrn

---

[7] von Baer to Dohrn
15/27 January 1873, Dorpat[30]

Dorpat, 15/27 January 1873

My dear Dr. Anton Dohrn,

Your most honored letter of 2 January unfortunately arrived during my absence. Since our year follows yours with a delay of 12 days, I was in the country celebrating Christmas with my children and grandchildren.

---

[30] Copy by Anton Dohrn; the original has not been preserved. While preparing to write the history of the Naples Station Dohrn preselected important documents from the Station's archives. These sources were arranged and listed according to years, persons, or subjects. The present letter from von Baer is listed as no. 6 of the 1873 correspondence with

On my return I found it; but since then through a chill during this winter journey, or maybe for another reason, I immediately fell into a serious sickness and am now not only out of bed, but also in fair recovery.

I thought I ought to begin with what I have just said to make you understand the delay of my answer. Passing to the contents of your letter, I must say, above all, that it has made an overwhelming impression on me. When quite some time ago I heard about your grandiose enterprise, I hardly believed that it would become a reality. I was still of this opinion when I had the honor of receiving your first letter. Since then, however, I have received so much public and private news about the splendid execution of your great project that I no longer have doubts about its completion, and believe that I recognize in it a new era for investigations in natural history. If on such a rich seabed as the Neapolitan the means for investigations are enhanced, and their methods improved and made available, successes must result that will be richer and more permanent than when, as was the case in former times, naturalists came there from far away with little preparation, limited apparatus, and without any knowledge of the sites, even without knowledge about the appropriate seasons. I therefore wish not to conceal that at the beginning I could not think of your enterprise without envy, since I had gone to Trieste almost 30 years ago,[31] to study the development of sea urchins, long before the spawning season of these animals, since there was no way to find out the correct time for such activity, and I had to return home just when the spawning began, because in Russia you have to predict when you will have finished.

I agree with you perfectly that there remains extremely much to do, in particular for embryology which illuminates not only zoology, but also understanding of animal life itself. Unfortunately my eye has abandoned me to such a degree that it no longer allows me to study the finer structure, and I have to be content that it shows me the direction in which to move about in my room. Nevertheless, I long to see your establishment during the coming fall if my state of health is then still tolerable, for the reminders increase that the clockwork, or better the structural support, of the body is worn out. I am close to my 81st birthday and that is a measure of time that already exceeds the upper limit of a normal lifespan.

You yourself, my dear Doctor, have furnished such thorough and excellent work in embryology that you need no recommendation to obtain the necessary governmental support for your establishment. According to

---

the annotation: "Very important letter! print *literally*!!" Dohrn copied the letter for the "appendix" to the manuscript of the history of the Institute (A. Dohrn, [Memoirs, see note 20], Appendix to pp. 95–96).

[31] In 1845 (July to September) von Baer had traveled for the first time to the Mediterranean, to Genoa, Venice and Trieste, with the intention of taking up again his embryological studies; this was followed by a second trip to Trieste from June 1846 to January 1847 (Introduction, p. 16; von Baer, 1866, p. 411; Raikov, 1968, pp. 208–213). In his article on the Naples Station (von Baer, 1873; see excerpt in Raikov, 1968, p. 212) von Baer talked in more detail about his frustrating experiences and working conditions while in Trieste.

my opinion, this establishment speaks sufficiently for itself. But I would be pleased if I could contribute in some way to bring this institution to life and to initiate its activity, although I would have liked even more to see it come into existence 50 years ago. First of all I am thinking of a public notice in the Petersburg German newspaper. But I would prefer to postpone this to the second half of February since I have reason to hope to be able to write again myself by that time. Until then, in our Nordic winter, the light is not sufficient for me to read and write. If until then you could send me your new discussion that you gave me hope for, I would be very pleased.

Heartiest thanks for your kind recollection of my earlier attempts, which I now consider only beginnings of a later comprehension. It is always pleasant when these first rudimentary attempts are not compared to the later more thorough ones.

With as much respect for your embryological works as admiration of your perseverance with the enterprise on which you embarked,

Your obedient servant
Dr. K. E. v. Baer

[8] Dohrn to von Baer
8 February 1873, Naples

Naples. Palazzo Torlonia. 8.2.1873

If what I have done here would deserve special acknowledgment, then the greatest I could ever receive would be what I found in your letter. That the Zoological Station has been linked through your letter directly to the great intellectual development in the center of which you stand, that thus this Institute, that is supposed to embrace firmly the forces of the present and the future and to work toward the solution of enormous problems, is still welcomed by the man who has taken up the traditions of Caspar Friedrich Wolff[32] and whose own history and development dates from the time of origin of all of modern biology—this has for me an everlasting value—, and the letter that brought this certainty to me is a patent of nobility which shall always remind me of *noblesse oblige,* should I ever incur the danger of turning my glance away from examples so high.

You must take my word that with regard to my merits, things are not exactly as you have put it so kindly. If fate allows a young man to live

[32] Caspar Friedrich Wolff (1734–1794), anatomist, member of the Imperial Academy of Sciences in St. Petersburg (1766); studied in particular individual development of animals, as well as malformations which led him to problems of inheritance. His work was highly influential through his emphasis on epigenesis in development.

according to his innermost drive–then this is probably only the merit of fate. In my case this becomes even more concrete, it was a pure and great service of my father, who gave me the means which alone enabled me to do what I have done. There is no need to add that there were difficulties at times that resulted from the novelty of the affair, from the peculiar situation of the City of Naples, and in particular from my personal inexperience. However, everything seems so happily overcome that I am fully confident that also in the future things will proceed well and fruitfully–in particular if all those who can best appraise the possible value of what may be achieved, do not deny their help to me.

I take the liberty of sending you a small pamphlet that has been published in August 1872 in the *Preussische Jahrbücher.*[33] It contains several things that may explain the scope and aim, reason and cause–in short the causality and teleology of the Zoological Station. In addition, I permit myself to send you the following further supplements that have been published in an article by Dr. Eisig[34] in the Viennese *Deutsche Zeitung;* its content originated with me.

Then I also add in this letter that after some extensive negotiations with the German Empire, the Prussian Minister of Education, and the Berlin Academy, Germany is ready to pay me a 10,000 Taler subsidy for the costs of building, that further, after the intervention of the Crown Prince, Prussia will rent *three* worktables in the laboratories of the Station for 1,500 Taler per year (500 Taler each), that negotiations have been started to have Saxony and Bavaria rent one worktable each and Italy two for the same amounts (500 Taler = 2,000 francs).[35]

Now I would like to turn to you, Excellency, in order to ask whether Russia, as well, does not want to acquire two worktables. Such a "table" is the collective definition for the supply of all those tools and provisions *except* the microscope which a zoologist needs at the seaside, i.e., chemicals, scalpels, needles, drawing and writing equipment, *aquaria for experiments, permanent provision* of living animals, eggs, larvae, extensive and very soon a complete *library.* Furthermore, and this certainly is very important, there will be an experienced scientific environment that [including personnel] will act encouragingly with help and advice,–in short providing a scientific existence that may rarely be found *outside* a university; it is often difficult enough to create inside a university and is even more difficult to preserve.

I also want to offer two tables to Austria-Hungary, one to Switzerland and if England, America, and other states or private persons want to

---

33 A. Dohrn, 1872.

34 Hugo Eisig (1847-1920), German zoologist and student of Ernst Haeckel, was the first assistant (1872-1909) and later the vice-director at the Naples Station; he also contributed to the series of monographs *Fauna and Flora of the Gulf of Naples,* published by the Zoological Station from 1880 on, by producing a monograph on *Die Capitelliden, nebst Untersuchungen zur vergleichenden Anatomie und Physiologie* (*Fauna and Flora,* 16; Berlin: Friedlaender, 1887).

35 At Dohrn's time 1 Taler = 4 francs = 3 Marks.

make such acquisitions, then there is still plentiful space to satisfy the needs. I shall also make the same offer to France but will certainly meet with a refusal, the more so since Monsieur Lacaze Duthiers[36] has already behaved very improperly on my behalf—which he is free to do in the future as well for his own pleasure.—

I wanted to tell your Excellency this and since you have been kind enough to promise your protection to the Zoological Station, you are now presented with the wish the fulfillment of which will benefit not only the Station but also others. May you succeed in preparing a good reception for it in Petersburg.

That we shall also have the honor to see you, Herr Geheimrath, here and to have you personally approve the new Institute, is almost more than I might have hoped for in the boldest of all moments. But if this should really come about I hope that your Excellency will let me know about it in time so that I can receive you at the very border of my new Fatherland, and so that I may ease as much as I can your traveling through the boot of Italy. In wishing for such an event I am joined by your special fellow-countryman Dr. Nicolaus Kleinenberg from Libau, the author of the recently published monograph on Hydra,[37] my faithful friend who preferred to develop the Zoological Station together with me and who therefore has left the German universities as I did. He asks me to give you particular expression of his respectful devotion to you.

And now I take my leave for today, hoping that this letter will find you in good health, and will bring to us and to the archives of the Zoological Station further communications in the old and honored master's very own hand.

Permit to the youth, Excellency, gratefully to squeeze the hand of the aged one.

Truthfully yours,
Anton Dohrn

---

[36] Henri de Lacaze-Duthiers (1821–1901), influential French zoologist, professor at the Sorbonne (1864), member of the *Académie des Sciences*, Paris, and founder of zoological stations at Roscoff (1872) and Banyuls-sur-mer (1881). Strong in nationalistic feeling, Lacaze-Duthiers expressed himself several times unfavorably against Dohrn, that "Prussian in Naples," and prevented French scientists for many years from going to Naples (Fischer, 1980; Groeben, 1985a, pp. 120, 226).

[37] Nikolaus Kleinenberg (1842–1897), born at Libau, Kurland, had studied medicine at Dorpat before changing to zoology and going to Jena, where he received his PhD with Ernst Haeckel. In 1872 Dohrn convinced him to join his Neapolitan venture. Three years later Kleinenberg, as strong-minded as Dohrn, left the Station and retired to Ischia. Due in part to an intervention of Dohrn, Kleinenberg was nominated to the chair of Zoology and Comparative Anatomy at Messina University in 1879 and in 1895 at Palermo. His work *Hydra. Eine anatomisch-entwicklungsgeschichtliche Untersuchung* (Leipzig, 1872) is a classic in zoology.

[9] von Baer to Dohrn
18 August 1873, Dorpat

Honored Doctor Dohrn,

I have been in St. Petersburg. There I found your letter to the Ministry;[38] it had been forwarded to the Academy of Sciences for an expert opinion. The Academy, however, is closed for holidays until 15 August old style (27 August new style) and therefore I found no academician present. I discussed the matter in detail with the Secretary of the Academy. At the Secretary's wish I also drafted an expert opinion in which I stressed in particular that the establishment would be of no greater importance for any other country than for Russia. Of all the seas that bathe European Russia, only the Arctic supports a variety of animals; on the Arctic, however, there is no city to be found, not even a well-illuminated house in which one could do microscopical investigations. The Baltic Sea east of Rügen is known to be more of an inland sea, with rather brackish water, the Black Sea is not much richer and the Caspian even poorer. We concluded that the Russian Empire should prepay for at least two tables and at the start the duration should be fixed for ten years. Only, since while looking through your application we found that Prussia and other states have subscribed for three years only, the Secretary felt that at the moment this period should not be exceeded. I have no doubts at all that the assembled academicians will insist on this proposition or propose more. Mr. Owsjannikow[39] has not yet returned, but will probably arrive at the end of August, and his vote will without doubt be of great consequence since the Russian government is not stingy with supporting the sciences, but gives probably more weight to the wishes of a native Russian than to those of a German.

At the end of our August or during the first days of September I plan to travel again to St. Petersburg and there I shall probably take part in these negotiations, because when I am there, I have again all the privileges of an academician. I tell you this in case you have something more to say about your establishment. In this case I would ask you to send me this postscript. Should I not be able to travel, which might well be with my very unstable health, I would mail this additional news. Before 12 September, according to your style, I shall find it hard to travel and I hope not to do so after 16 September.

The report of Dr. Eisig in the Vienna *Deutsche Zeitung* 72, no. 344, which you have been kind enough to communicate to me, I shall take

---

38 Dohrn to the Minister of Education, Count Tolstoi, 26.8.1873, Naples (A. Dohrn, Mailbook I; ASZN). In his letter Dohrn had included the first photographs of the barely finished building of the Zoological Station.

39 Philipp Vasilevitch Owsjannikow (1827–1906), professor of physiology and histology at the University of St. Petersburg, member of the Russian Imperial Academy of Sciences (1864).

with me to St. Petersburg. Perhaps you can also point out to me other, more recent reports.

With the heartfelt wish that all your enterprises that are so beneficial for science may prosper, I have the honor to sign as your

most sincere
Dr. K. E. v. Baer

Dorpat, 18 August 1873

---

[10] Dohrn to von Baer
7 September 1873, Brighton

Excellency,
I permit myself to offer you two photographs of the Zoological Palace in Naples[40] and I hope you might draw from it again the desire to pay us a visit down there in winter or at least next spring.

It has still needed great and unrelenting efforts to overcome the difficulties that have cropped up daily; and unfortunately there was also no lack of unforeseen accidents—such as the breaking of 10 big thick panes of glass worth 2,000 Taler.

The worst was that since January my energies started to slacken and the nervous depression reached such a degree that I had to retire completely from any work, and had to leave Naples in June because the heat became unbearable. Thus I have passed six weeks in Switzerland, 14 days with my sister in Baden, and since then three weeks in England at the sea to recover.

This has of course caused many new complications and the final completion of the Station will therefore probably be delayed until the end of the year.

May I bother you now, Herr Geheimrath, with a question? Months ago I asked the Russian government whether they would like to rent two worktables in the laboratories of the Station, following the example of Prussia, Italy, Bavaria, Baden, Alsace, Holland, and Cambridge University. The Moscow Society of Naturalists has strongly recommended the acceptance of this proposal; may I hope that you, Excellency, would recommend my formal request in a personal letter to Count Tolstoi?[41]

---

[40] In August/September 1873 Dohrn started to send the first photographs of the new building to his father (August 20), to colleagues (Emil Selenka, Leyden; Gustav Müller, Berlin; Prof. Harting, Utrecht) and to government offices (Prussian Ministry of Education, Berlin); see also note 38.

[41] Dimitri Andreevitch Count Tolstoi (1823–1889), Russian Minister of Education (1866–1880), president of the Academy of Sciences (1882–1889) and Minister of the Interior (1882–1889). Russia subscribed to a first table in April 1874 and to a second one in January 1875 (see Introduction, pp. 16–18).

FIGURE 2. The Zoological Station in 1873 (*ASZN*).

I know that things go slowly in Petersburg,—but just because of that it would certainly be very advantageous if the Minister is encouraged from your side as well.

And the Station urgently needs such support. Agassiz has received in one year a gift of 300,000 Taler,[42]—but in Germany American munificence is unknown. And I have always to struggle against scarcity of money.

And another request. There certainly exists a marble or plaster bust of Carl Ernst v. Baer; may the Station hope to obtain permission to set up a copy in the laboratory?

With these two questions and requests my importunity of today ends, for which I hope to obtain absolution, the more so since I committed them only in my role as founder of the Zoological Station and this office requires a certain importunity as *conditio sine quâ non* for its success.

Hoping that this letter will find you in good health and will earn me, through my father's address in Stettin, a few lines in reply I greet you and commend myself to you in due devotion and respect.

Yours truly,
Anton Dohrn

Brighton, 7 September 1873

---

[42] John Anderson, a wealthy New York tobacco merchant, had presented Louis Agassiz (see note 25) with a small Island, Penikese, off Cape Cod, Mass., and with $50,000 for establishing there a small summer school in marine biology. The "Anderson School of Natural History" opened for the summer of 1873 with 50 participants, most of them teachers. After the death of Louis Agassiz (1873) his son Alexander (1835–1910) took over the directorship for the second course (1874). Due to financial, technical, and personal reasons, and in spite of major interest among the participants, the school could not open for the third session (E. C. Agassiz, 1900, pp. 767–776; Lurie, 1960, pp. 380–381; 387; fn 40, p. 419; Maienschein, 1985, pp. 26–28).

[11] von Baer to Dohrn
31 August/12 September 1873, Dorpat[43]

To Dr. Anton Dohrn.

Dear Doctor Dohrn,

I have received last night your letter of 7 September 1873 from Brighton and I hasten to reply to it.

First of all I want to say that I have written to you about 12 days ago directly to Naples and I am surprised that the letter has not been forwarded to you. But perhaps it has not arrived at all since I did not know your correct address. I tormented myself to find a sufficient Italian address and wrote *Napoli nel nuovo Instituto* [*sic;* correct spelling is Istituto, Ed.] *Zoologico.* In this letter I communicated that I had been in St. Petersburg and found the business under way regarding the prepayment of two tables for Russian investigation. The Minister of Education, Count Tolstoi, had passed the proposal on to the Academy for an expert opinion; but the Academy was on holiday until mid-August, and only on 28 August according to our style could the proposal be discussed. Of course I declared my warm interest in its favor. The Secretary felt it would help if I would give public testimony. This I have done although, according to my opinion, those academicians who have recently been in Naples could have done so much better and the voice of a native Russian would have had a better effect with the Minister. However, being totally neutral, as I cannot enjoy the advantages of an Academician any longer– not only do my 82 years weigh heavily on me but also my eyes deny me their service–, my voice will at least be considered impartial.

I have therefore inserted in the Petersburg German newspaper an article which the Academy will call in testimony and which I enclose.[44] You are right that matters proceed slowly in Russia, in this case, however, it was predictable because the Minister of Education disposes of no funds that are not yet destined for something. He must therefore make a proposal to the Minister of Finance and ask for a special grant. To be able to do that he needs an expressive proposal from a competent authority, therefore the enquiry to the Academy. I do not doubt at all that it will be positive, but I have asked the Minister to let me know should there be difficulties. I would then write a second declaration, the contents of which is going to be: "какъ вамъ не стыдно" i.e., "you should be ashamed of yourself!", and I am sure such an expectoration from someone

[43] Dohrn copied this letter for publication in the appendix to his history of the Zoological Station (A. Dohrn, [Memoirs, see note 20], Appendix to pp. 96a–c).

[44] Karl Ernst von Baer, "Die Zoologische Station in Neapel." *St. Petersburger Zeitung,* no. 219, 1873. The article is dated 9 August 1873. No copy available. For a longer quotation see Raikov, 1968, pp. 212–213.

soon to be gone will not remain without effect. I have no doubts that before the end of this year two tables for observers from Russia will have been commissioned, but the order will probably only start with the coming year. I had already agreed with the Secretary that one should subscribe for 10 years, only when he found out that you yourself in your letter to the Minister had reported that Prussia and other states had subscribed for three years, he did not wish to go further.

You may trust me that it is more effective if I speak in public than if I write to Minister Tolstoi. He can throw such a document, but not a public declaration, into the waste basket. I am sure to carry enough weight that such a declaration will remain before him also in the future.

Heartfelt thanks for the photographs that represent a magnificent building. —But there exists no marble or plaster bust of K. E. v. Baer and it would now also not be very edifying because it could only show a decrepit person.

Finally I ask for an accurate address in Naples.

Sincerely yours,
Dr. K. E. v. Baer

Dorpat, 31 August (12 September) 1873

---

[12] Carl August Dohrn to von Baer 18 September 1873, Stettin

---

Stettin, 18 September 1873

Dear and honored friend and benefactor,

Many thanks for your welcome letter of September 12, —I have just returned from a trip to Carlsruhe to my daughter[45] and could not reply earlier.

Dr. Anton shall receive your letter which will certainly be a great pleasure for him, together with the insert (Renard[46] also sends me the Petersburger Zeitung) but I shall probably forward both items to England if, as I suppose, he is still there to be coached by Mr. Lloyd, Director of the Aquarium in the Crystal Palace, in the finer technica in the manipulation of the pipe system.

It would be a pleasure for me and of course for Anton if the Royal Society would feel pressed through your *vis prophetica* to fulfill your

---

45 Anna Wendt (1831-1892), married in 1857 to Gustav Wendt (1827-1912), classical scholar and dean of a gymnasium.

46 Karl Renard (1809-1886), physician, secretary of the Société des Naturalistes de Moscou.

prophecy. By the way, many and respectable societies have already sent brilliant items.

That *your* intercession for Anton will be of considerable influence not only for Russian but also for *other* Ministers of Finance is something of which I am deeply convinced.

Unfortunately one of Anton's patrons, Professor Czermak,[47] has just fallen victim to his several years' old diabetes. He was only 45 years old! What a pity for the excellent man. But since Patroklus also had to die–

But in the meantime we are going to watch the enterprise further, *diis faventibus*!

With longstanding devotion,
Your C. A. Dohrn

We expect the argonaut Heinrich back from New York during the next few days.

---

[13] Dohrn to von Baer
19 September 1873, London

London, Charing Cross Hotel, 19 September 1873

Excellency!

I hasten to add that Prussia is almost alone in subscribing for three years, the other governments subscribe for five, and I much prefer Russia to go beyond Prussia, because then I can push Prussia, which I shall do just now.

Holland has subscribed for one table, and Oxford is preparing to do the same.[48]

With greatest thanks for your participation and with due respect,

Sincerely yours,
Anton Dohrn

May I ask you to request for the Zoological Station all the biological publications of the Imperial Academy of Sciences.[49] The Academies of

---

[47] Johann Nepomuk Czermak (1828–1873), professor of physiology in Budapest, Jena, where Dohrn met him, and Prague. Czermak and his wife Marie Czermak-Lümel were among the friends who generously helped Dohrn with loans of considerable size to cover the building costs of the Zoological Station (1872–1873).

[48] It was only after several difficulties, and thanks to the personal intervention of Dohrn's friend, the British zoologist Edwin Ray Lankester (1847–1929), that in 1890 Oxford University also subscribed to a worktable.

[49] The St. Petersburg Academy of Sciences presented the first 18 volumes of its *Bulletin de l'Académie Impériale des Sciences* (vols. 1–18, 1860–1873) to the library of the Zoological Station (A. Dohrn, 1876, p. 86).

FIGURE 3. Anton Dohrn. Study in oil by Hans von Marées (1873) (*FA*).

Copenhagen (67 volumes), Berlin, and Naples have already contributed theirs. The Royal Society is about to do so and a great number of other, smaller academies and societies will then follow.

In great haste

Yours,
AD

I shall be back in Naples by mid-October.

---

[14] Dohrn to von Baer
29 September 1873, Stettin

Stettin, 29 September 1873

Excellency!

Yesterday I arrived here in my home town and hasten to thank you heartily for your second letter and for mailing the article in the St. Petersburger Zeitung.

Your first letter to Naples reached me shortly before my departure from London, together with another one that has been sent to me by a former member of the Department of Education in St. Petersburg and that contained the same good news as your letter.

With regard to the Station I can only say that it prospers remarkably, that all the innumerable difficulties are slowly but surely cleared away and that the whole affair has become so popular as has no other scientific enterprise for a long time.

During this last stay in England I have again succeeded in gaining several benefits for the new Institute. The English Society of Naturalists has granted me 200 Taler for performing experiments on the maintenance of eggs and larvae of lower marine animals; Oxford will, I hope, follow Cambridge, its sister university, and subscribe to a table, the Royal Society has donated a great part of their publications (your article in the St. Petersburger Zeitung has thus anticipated the truth) and a great number of other societies follows this example—the library grows *very* considerably and quickly.

Since the Academies of Berlin, Copenhagen, Naples etc. etc. have also donated their publications, I shall before long submit a similar request to the St. Petersburg Academy and hope not to meet with a negative answer.—

You will certainly be pleased to hear that *Huxley* has completely recovered. He was very sick, he suffers apparently from organic indigestion that has very much upset his nervous system. I visited him before my departure and found him in the midst of his family, as merry and cheerful as ever. He told me to give you his heartiest greetings.—

My father wishes you untroubled continuance of your marvelous health–no need to add that I too, and with me the whole company of those who are affiliated with the Zoological Station join in this wish.

I hope to be able next year to report that work in the laboratories has started–a great number of German, English, Italian, Dutch zoologists have already made reservations.–

With all due devotion and most sincere thanks

Yours,
Anton Dohrn

My address in Naples is the following:
Dr. A. D., Stazione Zoologica

---

[15] von Baer to Dohrn
18 October 1873, Dorpat

Dorpat, 18 October 1873

My dear Doctor Dohrn,

I shall not omit announcing to you that the Academy of St. Petersburg has applied for the support of two tables. It is not to be doubted that the Czar will make the grants, if only the Minister forwards the application; but I still remain with my supposition they will start only next year. For my part I have also insisted to the Academy that they should send you their biological publications, although I hardly believe that that was necessary, in particular if you have written to them that so many other academies have preceded them. By the way, it has been some time since I have had more recent news from St. Petersburg.

With the heartiest wish that your health may be reestablished permanently, which is so necessary for the growth of the new establishment, I sign with as much cordiality as respect

Your most obedient servant,
Dr. K. E. v. Baer

[16] Dohrn to von Baer
23 January 1874, Naples

Stazione Zoologica di Napoli
23 January 1874

Excellency!

"But all remains silent as before!"[50]

Count Tolstoi has not mentioned a word as to whether he is inclined or disinclined to accept my offer and to consider the recommendation of the Academy as more than just valuable material. The Academy may settle this with Count Tolstoi as it likes,—that is its business, but as for myself I think it most necessary to face him with an either-or, since requests have been forwarded to the Zoological Station from other directions that I have to decline for as long as I have to keep the two tables at the disposal of the Russian government.

But before I send a letter in that sense to the Minister—and the Count seems not to be aware of the fact that I am an independent fellow and well worth the courtesy of an answer—, I wanted to communicate first to you, since you have looked after my enterprise with so much interest. Perhaps you can advise me what would be the best thing to do to avoid spoiling the game.—

In spite of untiring activity I am still not finished with all the installations. This is partly the fault of a wearisome nervous prostration resulting from a tension that has lasted too long, together with mental emotions, and also the result of the complexity of the task.

Nevertheless the aquarium, which contains 60 tanks, will be opened in three days and I hope the public will like it.

For next month a Swede and three Englishmen, and for March two Dutchmen have been scheduled,—there is therefore no lack of people anxious to work for whom the Station will be of advantage.

And at the end of these short items of communication I have to confess that we have in spite of all set up your bust in the ceremonial room of the Zoological Station. It has been made by the young sculptor Adolf Hildebrand,[51] who recently has acquired a considerable reputation,

---

[50] A quotation from Goethe.

[51] The German sculptor and architect Adolf von Hildebrand (1847–1921), a friend of Anton Dohrn's from Jena times, had already helped Dohrn with the classical façade of the Zoological Station (1872) and would later (1885, 1901) give advice concerning the additional buildings. In spring 1873, on the occasion of the World Exhibit in Vienna, Hildebrand had exhibited two statues that attracted much attention. During the summer of the same year (1873) the German painter Hans von Marées (1837–1887) and Hildebrand created a cycle of frescoes for the ceremonial room of the Zoological Station, originally destined to be used for concerts and entertainment. For this room Hildebrand also created two more than life-size busts of Karl Ernst von Baer and Charles Darwin, both of whom Dohrn considered the fathers of modern biology. Hildebrand took as a model for the statue of von Baer the reduced steel engraving of a drawing by P. Borell, published on the frontispiece of von

FIGURE 4. Karl Ernst von Baer, bust (plaster) by Adolf von Hildebrand, 1873 (*Stazione Zoologica "Anton Dohrn"*).

from the portrait in your autobiography and it resembles you very much, as someone told me who has frequently seen you in Dorpat. Will you also approve this attempt?

Hoping that in spite of the winter you still have your old vigor, I greet you with all my heart, and I have been charged by my friends and collaborators at the Station, Dr. Kleinenberg and Dr. Eisig, also to present their most respectful compliments.

My father, who is in Genoa, and who next month will visit the building and all that it contains, will I hope find that his son has spent his money to good advantage. If he knew that I would write to you, a greeting in his style would surely not be missing.

Yours respectfully devoted,
Anton Dohrn

---

[17] von Baer to Dohrn
24 January/5 February 1874, Dorpat

My dear Dr. Dohrn!

I received your favor of 23 January a few hours ago and I hasten to dictate an answer to it.—

Count Tolstoi, who tries his hardest to enhance the study of ancient languages in Russia, although the Russians have up to now shown not much liking for this study, may believe that it is not necessary to promote the study of nature because as Russians they have an already inborn taste and talent for that, or he may just have dawdled time away; in any case he has acted very unmannerly by keeping silent for so long. It seems quite urgent to me to ask him for a categoric answer since it is out of the question that you reject others to wait for him. I believe you could write roughly the following:

> You had proposed to the Russian Empire to subscribe for two tables, convinced in this way to do the Empire a favor, and to prove your respect. Since during the last years some young naturalists of Russian nationality have contributed well-esteemed and excellent studies on the embryology of marine animals, and since Naples was frequently the site of their investigation, you had felt that the conveniences and means that your Institution provides should also

---

Baer's *Autobiography* (von Baer, 1866). The drawing, which had been done in 1864 for the 50th jubilee of von Baer's doctoral degree, shows von Baer at the age of 72 (Knorre, 1975, pp. 243–244, Fig. 10). The bust shows von Baer without glasses and with a slightly different hair arrangement. Knorre (1975, Fig. 17, p. 252) assumes that more than one photograph had served as a model, but at that point Dohrn did not yet have any portrait photograph of von Baer, who sent three photographs of himself to Dohrn in August and December of 1874 only (see letters 23, 26 and Figs. 5–7).

be offered to Russians. But you had not yet received an answer. But claims have now been made from other directions for the tables that are still unoccupied and you had no possibility of rejecting them as long as you have not received an assurance from Russia. I would even fix a deadline up to which you are willing to wait, or after which you will give up seeing Russian naturalists in your Institute.

I believe you will gain for the future if you do not come asking for something since one expects a categorical declaration.

It is really remarkable that young Russians have distinguished themselves in this branch of finer observation although their sea shores are not at all inviting for such work, and that at the moment the government does not seem to want to do anything at all for these studies and instead forces upon them other disciplines for which Russians have no taste at all.

I shall also write to the Secretary of the Academy to let him know what effect their recommendation has had. Just a few days ago a local naturalist told me that he would like to go to Naples to revise the development of Amphioxus.

So you have indeed set up my bust; marvelous enough, but too much of an honor for me. If I still hold together sufficiently this summer, I still do wish to see your establishment and then to take leave of the world. However, reminders of friend "Hain"[52] become more and more serious and at the beginning of this year I have been seriously ill and very weak. It depends entirely on how the transition to our spring goes. Now we finally have winter which has been slow in coming.

Recently I have received from Agassiz detailed news about the funds that have been put at his disposal for his museum. He still on his deathbed had them accumulated.[53] I am quite amazed about the munificence of these funds. What a difference between America and Europe!

Last summer I published a treatise in which I express my doubts on the interpretation that has been given to the development of ascidians.[54] Since this treatise has been published in the Memoirs of the St. Petersburg Academy it seemed to me superfluous to send it separately; but since the Academy also often pays homage to the principle: "What takes a long time will be good," I shall try to mail an individual reprint separately.

---

52 Poetic euphemism for death.

53 Agassiz had died on 14 December 1873. On 13 August 1873 he had written to Dohrn from Penikese: "I believe I have not told you yet that on my birthday, 28 May, my son-in-law, who exploits significant copper-mines at Lake Superior, has presented me with a gift of *one hundred thousand* dollars for the enlargement of my museum which I will mainly use for expanding the collections, hoping that the State will take care of the enlargement of the building that last year we have already enlarged to about twice its size. And mainly I want to begin to bring authentic collections over from Europe in order to implant tradition that is missing here so much" (*FA: Ba 659*; translated from the German).

54 K. E. von Baer, "Entwickelt sich die Larve der einfachen Ascidien in der ersten Zeit nach dem Typus der Wirbelthiere?" *Mém. Acad. sci. St. Petersb. (sci. math. phys. nat. Kl.)*, (7. sér.) 19 (1873), no. 8: 1–35.

With heartfelt congratulations for the progress of your enterprise I have the honor to sign as your

Most respectful
Dr. K. E. v. Baer

Dorpat, 24 January/5 February 1874

P.S. May I ask you to convey my cordial respects to Mr. Kleinenberg, whose excellent publication on Hydra I know very well.

B.

---

[18] von Baer to Dohrn 23 March/5 April 1874, Dorpat

Dorpat, 23 March/5 April 1874

My dear Dr. Dohrn!

Much to my deep regret I have to tell you that occupation of only one table of your establishment may probably be expected from the Russian government. The enclosed letter from the permanent Secretary of the Academy of Sciences of St. Petersburg,[55] addressed to me, will give you more detailed information. I had asked the Secretary to see to it that two tables would be reserved; the enclosed letter contains his answer. It is almost incredible that the Minister of Education is so lukewarm in supporting a line of studies in which native Russians have distinguished themselves several times, and that he wishes to nationalize instead so-called "classical studies" in which no Russian has ever distinguished himself. It seems to me that the whole direction is contrary to Russian intellect. –I shall probably publish another short article,–but it will probably be in vain.

Did you receive my answer to your last letter and the essay on "the ascidians"? I do not ask without reason. Unfortunately it transpired that recently several mailings of letters and revenue stamps that had been mailed from Dorpat have not been received. A "stamp collector" is suspected who could be either here or in Riga.–

I still hope that the subscription to two tables may still be achieved for the New Year. But I ask you to tell me occasionally what kind of privileges, book mailings and the like, you have received from other directions. In Russia one loves to follow examples of others.

Respectfully,
Yours sincerely,
Dr. K. E. v. Baer

---

[55] Not preserved.

[19] Dohrn to von Baer
16 April 1874, Naples

Napoli, Stazione Zoologica, 16 April 1874

Honored dear Geheimrath!

Since yesterday I am back in Naples after my doctor had advised me to go to Sorrento because my health had not improved during the winter, but rather had deteriorated, and for the last six weeks I had, in addition to my persisting nervous complaint, bronchitis and attacks of malaria.

All this has cut me off almost completely from my activity and also from correspondence and I therefore have not only to thank you for your untiring help and interest but also to ask your forgiveness for the delay of these thanks and of my reply.

I received from the Russian government the announcement that *one* table is to be taken.[56] I still have to reply to this letter as well. In the meantime I also received from other directions elucidations of the whole matter and I have reason to believe that rivalries between elements from Moscow and from St. Petersburg were involved. But I have been assured from all sides that at a renewed request the second table will probably also be subscribed to. And that is of course very desirable.

Permit me to give you the following information about the state of the whole affair.

The costs of the whole enterprise have grown in such a way that my own means lagged far behind the need, and I therefore felt obliged to borrow money from friends.[57] That was bad enough. But at the same time the profits of the aquarium have also remained lower than expected. Therefore during all of this winter I had to face an extremely difficult situation, the effects of which have obviously also been disastrous for my nervous state. I had already been prepared to experience a catastrophe when help came from England. Huxley, with whom I have been for years on very intimate terms, suggested the idea of a subscription among English naturalists and amateurs in such fields,[58] and the first who signed up was Darwin. Together with an extremely friendly and delicate letter he sent 120 pounds sterling as the contributions of himself and both of his sons. It is hoped that 1,000 pounds sterling will be collected in this way.

I hope to be able to start a similar procedure also in Germany and I

---

[56] Ministry of Public Education to Dohrn, 22.2./6.3.1874, St. Petersburg (*ASZN: G.LVI.3*).

[57] Among the friends who helped Dohrn with gifts and loans were included the art critic and patron of Hans von Marées, Conrad Fiedler (1841-1895), the British embryologist Francis Maitland Balfour (1851-1882), his brother Gerald (1853-1945), the physiologist Johann Nepomuk Czermak (see note 47), Dohrn's father-in-law Georg von Baranowski (1821-1914), and the German Consul in Naples, Otto Beer (1824-1891).

[58] Huxley had heard from F. M. Balfour, who worked as a guest investigator at Naples from February to June 1874, that Dohrn lacked funds for equipping the laboratories. The goal of £1,000 was reached in a few months only.

shall even turn directly to the crown prince in Berlin to ask him to place his name at the top of such a list.[59] Otherwise I am afraid I shall receive very little from wealthy Germans. At the same time, I shall try again to ask for another subvention from the German Empire.

In this way I hope to free the young Institute of debts and at the same time to bury it still deeper in the public interest.

I would of course prefer, if possible, to raise the annual contributions of the governments to such a degree that in case of need they could cover the operating expenses if during a war or cholera epidemic tourists would stay away from Naples. But whether this can be done remains doubtful to me.

At present three Englishmen, two Russians, two Germans, two Italians, and one Dutchman work at the Station. In addition there are 5 or 6 other Russian zoologists in Naples and one Swiss in Messina[60]—sufficient evidence of how much visits have increased and how useful it is gradually to turn the Station into a center whence help and support for everybody can originate.

But the governments should tell themselves that this enterprise can really reach this goal only if there are available sufficient funds for safe life,—and this is why it is so important that fixed annual contributions come in. Prussia pays 1,000 Taler but has not sent a zoologist,—but Minister Falk[61] has assured me that this annual contribution cannot any longer be cancelled—nor can Bavaria's 500 Taler.—

I did indeed receive with heartfelt thanks your work on the ascidians and have read it immediately. But at that time I was not able to answer immediately.

This big question does occupy all zoologists at the moment and although it may be difficult to foresee which direction is pointing toward the solution, it seems as if we will not have to wait much longer for an answer.[62]

My own position in this question is quite an isolated one,—it refers to Geoffroy's old definition of the vertebrates proceeding on their backs, —the insects. Through an egg of Columbus I believe to be able to provide

---

[59] In spite of several attempts and drafts (1874/75) a similar subscription took place in Germany only a few years later. R. Leuckart, R. Virchow, E. du Bois-Reymond and C. T. von Siebold had declared their willingness to sign such a document. Dohrn had hoped to collect funds for an urgently needed steamship, which, however, was given to him by the Berlin Academy of Science and the Prussian Ministry of Education (1876–1877).

[60] No details are available on the zoologists mentioned here. Besides Trieste (see also note 31), Nice and Messina, this above all in the wake of Johannes Müller (1801–1858), Naples also attracted many zoologists for seaside studies, even before the opening of the Naples Station. Among them were included Albert von Kölliker (1852), I. I. Metchnikoff (1865–1868), Alexander Kowalewski (1865, 1868, 1870) and also W. W. Salensky (1868, see note 64) (I. Müller, 1976, pp. 75–76). A detailed study of working conditions and facilities in these other localities, of local traditions and the transmission of information among scientists would be interesting.

[61] Adalbert Falk (1827–1900), Prussian minister of intellectual, educational and medical affairs (1872–1879).

[62] This question still remains open in 1992.

here a satisfactory answer to it, – but it will still entail so eminently much observation that, now so completely separated from my own research by the Station, I cannot, at the moment at least, think of interfering in this dispute.

During the summer I want to conclude my embryology of the insects, and then the state of the Zoological Station will I hope permit me during the winter to further in Naples my investigations on fish embryology started many years ago. In the meantime so many investigations on Ascidia, Salpa, Myxina, Petromyzon etc. will appear that there will indeed be a large amount of material on which to base conclusions that will allow the verification of eventual deductions. May they proceed from all sides *sine irâ et studio*! –

Will you find it immodest, Excellency, if I enclose a photograph in this letter and timidly ask whether there is hope of obtaining for myself a portrait of you? I would have liked to try to present myself to you personally, but although I shall come in about four weeks to Warsaw, where I am going to fetch my bride,[63] who has been engaged to me for almost one year and a half, I may at the moment not think of anything else, but following my doctor's strict instructions, I shall go directly to my father's country house and pass the summer there as quietly as possible. During my absence Dr. Kleinenberg and Dr. Eisig will manage the Zoological Station as they already do now during my illness. –

Your kind inquiry regarding the mailing of books etc. for the benefit of the Station I can so far answer as follows: above all the Royal Society – again through Huxley's mediation – has stood the test most splendidly. But also other Academies – Copenhagen, Amsterdam, Naples, Vienna, Berlin, Smithsonian Institution, have donated much. The Petersburg Academy has promised to do so, but up to now I have not yet received anything. I am particularly anxious to receive the publications of all Russian universities regarding biology. I plan gradually to request this. At present however I have to concentrate all my efforts on providing more financial means for the equipment of laboratories with instruments and certain conveniences.

Whether these efforts will lead to success remains to be seen, but with your permission I shall take the liberty of occasionally reporting on them to you since you have taken the Zoological Station under your protection.

And I therefore express my most heartfelt thanks to you for the manifold proofs of this beneficial protection which honors us, and I wish with

---

[63] On 3 June 1874 Dohrn married Marie von Baranowska (1856–1918) in Warsaw. Dohrn had met the Polish-Russian family von Baranowski at Messina in 1868. The father, Georg Ivanovitch (1821–1914), had to leave state service in 1863 (he had been governor of two Russian provinces) when during the Polish uprising his wife Catharina (née von Timler, d. 1875) had openly shown her sympathy for her people. After the father's return to Russia the mother, together with her two daughters Marie and Helene, moved to Naples; there Dohrn met them again in October 1871. Anton and Marie Dohrn had five children: Boguslav (1875–1960), Catharina (1876–1877), Wolfgang (1878–1914), Reinhard (1880–1962) and Harald (1885–1945).

all my heart that health and participation in matters of this world may be at your command still for a long time!

As to Count Tolstoi, I shall downright besiege him until he capitulates with the second table—and for this I count on the help of the Russians who are here at the moment.[64]

Hoping to obtain your pardon for the long delay of this answer, which has not been my fault, I remain

Respectfully yours,
Anton Dohrn

In four weeks I plan to be in Stettin in my father's home.

---

[20] Dohrn to von Baer
28 May/[9 June 1874], Petersburg, telegram

Dorpat, Academician Carl Ernst von Baer.

Are you at home these days—answer prepaid.

Anton Dohrn

---

[21] von Baer to Dohrn
28 May/[9 June 1874], Dorpat, telegram

I am at home.

Doctor K. von Baer

Dorpat, 28 May

[64] The following two Russian investigators worked at the Zoological Station at that time: Dr. Rajewsky from Moscow and W. W. Salensky from Kasan; the latter returned again in 1880, 1891, and 1907. Regarding the second Russian worktable see note 41 and full discussion in Introduction, pp. 16–18.

[22] Dohrn to von Baer
11 June 1874, St. Petersburg

M B
St. Petersburg. Fantanka 13.
Feodoroff home.
11 June 1874

Excellency!

In spite of all my efforts to settle matters here quickly, in order to be able to pay you a visit and to thank you personally for all you have done in supporting my efforts, I realize that I have to give up this intention, which I greatly regret.

It is the Zoological Station itself that forces me to make this renunciation–and my only consolation is that I am successful.

You will be pleased to hear that I have succeeded in stirring the heart of the Minister of Finance Reutern[65] and in moving him also to open a credit for a second table, similarly for 5 years. On Count Tolstoy's [*sic*] side this had been requested from the beginning, but it encountered opposition from Mr. Reutern which I have fortunately eliminated.

But *l'Appetit vient en mangeant,*–at present I am still on my knees trying to persuade the Medical Academy to acquire a third table. My government recommended me to Prince Reuss,[66] who then recommended me to the Minister of War who will receive me in audience before long. I hope this plan will be successful.

At the beginning of next week I go from here to Stockholm and Copenhagen on the same business, although with less hope for success.–

The English subscription, of which I wrote in my last letter, has up to now yielded between 800 to 900 pounds sterling,–a tremendous help! In Berlin I have furthermore tried to diminish my deficit and hope to receive this October 5,000 Taler from the Prussian government and next October the same sum from the German government. If the Minister of War Milutine [*sic*][67] approves, 14 tables will be subscribed for and this represents an income of 7,000 Taler–with this the original plan approaches completion, namely: the gain of disposable surpluses, which will allow the Zoological Station to hire younger scientists for the careful working up of the fauna of the Gulf,–and to investigate in particular the physico-chemical, geological, and botanical conditions of life of this fauna and to so to speak evaluate it statistically. A distant goal,–but still no Utopian idea.–

---

65 Michael Christoforowitch von Reutern (1820-1890), Russian Minister of Finances (1862-1878).

66 Heinrich VII, Prince Reuss (1825-1906), German envoy, later ambassador, in St. Petersburg (1867-1877).

67 Count Dimitri Alekseevitch Miljutin (1816-1912), Russian statesman, Minister of War (1862-1881).

May I, dear Dr. von Baer, report to you in all modesty on another matter that has already interested me for some time, and of which I have spoken in a small publication which I believe I have also sent to you in its time. Through our studies on the beginnings of embryology in its historical development my friend Kleinenberg and I have gotten more and more interested in the tremendous achievements of Caspar Friedrich Wolff, and we both have regretted that the Petersburg Academy has not yet published all the writings of the great man that are said to exist in manuscript. Since you, Excellency, have once had this matter in your control, I take the liberty of asking whether perhaps there exists an obstacle that might possibly be overcome. Unfortunately most of the academicians are already traveling so that I have to direct my enquiry to you directly. Could the Academy perhaps be influenced to publish a complete edition of Wolff's writings? Or would one entrust such a task perhaps – in case no greater authority takes up the matter –, to Dr. Kleinenberg, who is German-Russian by birth, and myself, a fellow-countryman of Wolff? We would both be very proud to use every effort to carry out such a task of reverence, which at the same time would be of such high scientific and historical significance. But of course, at first it would have to be possible at all. –

Mr. von Bradke[68] in the Ministry of Public Enlightenment has charged me to transmit his respects to Your Excellency. Unfortunately I have to give them in writing. At the same time he told me that Dr. Rosenberg[69] has been awarded the use of the first of the two Russian tables, so that he can write to Dr. Kleinenberg, his fellow-student, as soon as he wishes to settle all the details. I would have liked very much to talk to Dr. Rosenberg; but fate and General Milutine do not want this.

Since I stay here only until Tuesday at best there is hardly a chance to receive from you even a short note only, Excellency, – however, should this be possible, I would be very pleased, in particular if you would add that you are well and that your health holds up in spite of the bad weather.

With this heartfelt wish and the plea to accept in friendship my repeated requests, I am,

Yours truly and thankfully,
Anton Dohrn

Please give my friendly regards to Dr. Rosenberg.

---

68 Emanuel von Bradke (1832-1918), head of department in the Russian Ministry of Public Enlightenment.

69 Emil Rosenberg (1842-1925), professor of comparative anatomy, embryology and histology at Dorpat, professor of Anatomy at Utrecht. Rosenberg worked at the Naples Station from 7 April to 20 June 1875 at one of the two Russian worktables.

FIGURE 5. **Karl Ernst von Baer at the age of 67, photograph by Ph. Hoff, Frankfurt a.M. (1859)** (*ASZN: La 6*).

[23] von Baer to Dohrn
25 July/6 August 1874, Dorpat

To Dr. Anton Dohrn

Dorpat, 25 July/6 August 1874

My dear Doctor!

I have regretted very much that circumstances have not permitted you to visit me in Dorpat. If only I had known that you would travel to St. Petersburg, I would have gone there, since I had to settle a few things there anyhow. But your arrival here had been announced to me provisionally by a lady, and while I was hoping for your visit I received the telegram from St. Petersburg that fostered the same hope, and some time later the letter that announced your departure, the latter however arrived only on the same day that you had fixed for your departure. There was therefore no possibility of answering it. Later I myself have been away from Dorpat.

I am very glad that your wishes have been fulfilled in St. Petersburg. I could not have advised you anything better than turning directly to the respective offices, because in Russia one is more obliging and considerate to foreigners than to natives.

I still have to humor your wish by which you have honored me some time ago by asking for my photograph. I may assume that you were thinking of a photograph the size of visiting cards; but even then some doubts remained whether I should send you my most recent photograph

FIGURE 6. **Karl Ernst von Baer at the age of 80 (?), photograph by Th. John, formerly Schlater, Dorpat (1872?) (*ASZN: La 5*).**

or an earlier one, since the former shows me only as decrepit. A graduate of life commends himself less than a graduate of a school or a university. I therefore wanted to leave the choice to you between the last photograph and one made in Frankfurt fifteen years ago. This could have easily been taken care of during a personal encounter; now the only way out is to send you both and to leave the choice to you.[70]

I assume that you are still at Stettin with your father and I therefore send this letter there.

Respectfully,
Dr. K. E. v. Baer

---

[70] K.E. von Baer at the age of 67, photograph by Ph. Hoff, Frankfurt a.M. (1859), see Fig. 5 (*ASZN: La 6*). Knorre (1975, p. 240, Fig. 7) dates this photograph "around 1860, photographer unknown."

K.E. von Baer at the age of 80(?), photograph by Th. John formerly Schlater, Dorpat (1872?), see Fig. 6 (*ASZN: La 5*). See also Knorre, 1975, p. 252, Fig. 16.

[24] von Baer to Dohrn
26 July/7 August 1874, Dorpat

To Dr. Anton Dohrn

Dorpat, 26 July/7 August 1874

Honored Doctor!

Yesterday, I had just sent the letter to you to the post office when I remembered that there was one point that I had not answered at all, namely the question regarding the literary legacy of Casp. Fried. Wolff. I ask you therefore to consider this sheet as a postscript to my earlier letter.

There exists, as far as I know, only one work left by K. F. Wolff. This work only deals with double monsters and only with those of man. It is very specialized and accompanied by a great number of neatly made drawings. I had once promised to see to the publication of this bequest.[71] But I shied away from the great costs which this quantity of drawings would entail. In particular, the gain in knowledge seemed to me low in relation to these costs, since the means of the Academy unfortunately are very limited. The main results of the investigations on double monsters, viz. that the designations of two so-called conjoined individuals agree, have already been obtained and have taught that they are not fusions but separations. The Academy has certainly behaved irresponsibly in not having published the work earlier; but it remains very questionable whether it should even now spend money on it. If the figures are done abroad they will not only be less expensive but also will be done better. That you, dear Doctor Dohrn, and your excellent assistant, Dr. Kleinenberg, are well qualified to publish this work, is self-evident and will certainly not be doubted by the Academy. If you believe it can be published without the assistance of the Academy, the latter will certainly accept such an offer with pleasure. However, should the Academy have to bear the costs, I would have doubts about their decision. There are always more works offered to the Academy, most of them with numerous illustrations, than it can bring to light. Toward what its small means should be allotted depends on the general conference, i.e., the consent of all the Academicians.

Respectfully,
Dr. K. E. v. Baer

---

[71] Over a longer period (1830-1846), von Baer had repeatedly worked on the cataloguing of the literary legacy of C. F. Wolff (see note 32) in the archives of the Academy; he had already reported in detail on this in: "Ueber den litterärischen Nachlass von Caspar Friedrich Wolff. Erster Bericht," *Bull. math. Acad. Sci. Petersb.* 5 (1847): 129–160. Wolff's unfinished manuscript has only recently been published: K. F. Wolff, *Reflections in connection with a theory of malformations.* Leningrad: Nauka, 1973. See also: A. E. Gaissinovich, "The Development of Embryological Research in the Academy of Sciences of the USSR (on the two hundred and fiftieth Anniversary of the Foundation of the Academy)," *Soviet J. Developmental Biol.* 5 (1975): 185–192.

[25] Dohrn/CAD to von Baer
10 August 1874, Stettin

Stettin, 10 August 1874

Excellency!

You have given me extraordinary pleasure by your letter and the two accompanying photographs. But how could I have the heart to return one of them? We are all students; and I am afraid the results of my life-examination will not be so splendid as to entitle me to the eighties, in particular since the *locus minoris resistentiae* has already protested during recent years and has called attention to itself.

I never tire of expressing my admiration to a man of advanced years who has solved the problem of living pleasantly with greatness, and for a long time, and I am doubly distressed to have been forced to pass by Dorpat without having been able to express this admiration personally with all modesty.

Looking back, I have to regret this even more because of the reason for my longer stay in St. Petersburg, viz. the third table to be supported by the Medical Academy has fallen through, unfortunately into the waters of the Neva and not the Gulf of Naples. At least I have heard that in spite of a peremptory recommendation of the Minister of War–whose letter to Prince Reuss I have–the Academy has rejected the table. I shall once more take direct steps with General Miljutin, to whom I have very good introductions,–but I am afraid to have to deal with intrigues, or with sensitivities; to eliminate these will be difficult if not impossible.

At present I have the pleasure of reading your studies on the rivers of Russia;[72] I have also read the protest against teleophobia[73] repeatedly and, as I may honestly say, with much inner satisfaction. I openly admit to you, Excellency, that I am decidedly a disciple of the Darwinian theory and that I am convinced that the demands of nobler intellects may be satisfied by the deepening and further development of the great idea. The Huron howlings[74] that emanate from certain directions disgust me deeply–but unfortunately in our brutal times it finds too much fertile soil–and too many editions. But just because I have subscribed heart and soul to the ideas of Darwin, it annoys me doubly to have to observe this wild activity. At present, however, my hands are tied for a hundred reasons, and as a pure Darwinist I cannot, in spite of my innermost

[72] K.E. von Baer, "Ueber ein allgemeines Gesetz in der Gestaltung der Flussbetten" (Kaspische Studien, 8). *Bull. Acad. Sci. Petersb.* 2 (1960): col. 1-49, 218-250, 353-382; addenda: *Bull. Acad. Sci. Petersb.* 7 (1864) col. 311-320. There, von Baer published for the first time in German the law formulated by him and later named after him (Baer's Law) (Raikov, 1968, pp. 254-255).

[73] On von Baer's attitude with regard to teleophobia see: Raikov, 1968, pp. 404-409.

[74] Probably an allusion to Ernst Haeckel, whose popularization and philosophical interpretation of Darwin's ideas Dohrn would approve less and less.

wishes, let off steam against those other impure Darwinians. The way in which David Strauss[75] has played into the hands of this nonsense was just what was needed to improve this St. Vitus' dance and it cannot be foreseen where else the wild chase will end up. I am thus the more anxious to see your promised work.

With the plea to replace the photograph sent to you earlier by the enclosed one, which according to my family is much better, and to permit me the satisfaction of inserting *two* portraits of Carl Ernst von Baer into my album, I send you my greetings and my best compliments, and request the continuance of your interest in my rope dancer's enterprise in Naples.

Yours respectfully and devotedly,
Anton Dohrn

[CAD:]
The commissioner, who for the above reasons finds the commission to forward the letter easy enough, readily takes the welcome occasion to add his heartiest greeting and that of Dr. Heinrich, who has just left for the train. Since I just permit myself the byplay of translating a piece of Calderon, the following Spanish wish seems to be befitting

Viva V[uestra]m[erce]d mil años
Your C. A. Dohrn

At my *tusculum*
11 August 1874

---

[26] von Baer to Dohrn [received: 10 January 1875, Dorpat]

Honored Doctor!

To present you with the most recent example of a visible transformation I send you the enclosed photograph.[76] It is the most recent portrait of Dr. K. E. von Baer, known to you, who has for some years grown a beard and who has therefore been urged to sit once more for a photographer. The man looks very ill-humored, no doubt because he can no longer use his eyes.

---

[75] In his last publication, *Der alte und der neue Glaube* (Bonn, 1872), the German philosopher and theologian David Friedrich Strauss (1809-1874) had confronted Christians with an evolutionary view of the world. The book was a tremendous success, undergoing two further editions during its first year, with nine more to follow before 1881.

[76] K. E. von Baer at the age of 82, photograph by Th. John, formerly Schlater, Dorpat, end of 1874, signed "Dr. K. E. v. Baer", see Fig. 7 (*ASZN: La* 4). Knorre (1975, p. 255, Fig. 19) gives 1876 as the date of this photograph.

FIGURE 7. **Karl Ernst von Baer at the age of 82, photograph by Th. John, formerly Schlater, Dorpat (end of 1874), signed "Dr. K. E. v. Baer" (*ASZN: La 4*).**

Together with the *Zeitschrift für wissenschaftliche Zoologie* I have just received the catalogue of the library of your Institution.[77] This library is already very rich. One could sit for a whole year in Naples to orientate himself in the zoological literature alone, if one were lucky enough still to be able to use one's eyes.

I have recently written an article on Darwinism[78] which is supposed to go into the second volume of my speeches if the publisher wishes to print it. I have become half a Darwinist, or rather a transformist; but I

---

[77] A. Dohrn, 1874. The publication was sent as an enclosure to all subscribers to the *Zeitschrift für wissenschaftliche Zoologie.*

[78] K.E. von Baer, *Ueber Darwin's Lehre.* In: *Reden gehalten auf wissenschaftlichen Versammlungen und kleinere Aufsätze vermischten Inhalts.* Vol. 2: *Studien aus dem Gebiete der Naturwissenschaften.* St. Petersburg: H. Schmitzdorff, 1876, pp. 235–480.

can become neither a full adherent nor a full opponent. I have been left behind on the old sandbank which bears the heading: Nescimus. But just for that reason I have no objection when others who have more hope undertake their investigations from the Cape of Good Hope. Even if they should not reach the South Pole they can nevertheless get closer to it. But I shall not give up the idea that all development strives toward a purpose because without a purpose I cannot imagine development. Perhaps out of Naples will appear someone who can discover the mechanisms of development of which I spoke on the last page of the preface of my *Entwickelungsgeschichte.*[79]

With heartfelt wishes for the New Year,
Yours respectfully,
Dr. K. E. v. Baer

---

[27] Dohrn to von Baer
17 January 1875, Naples

Naples, Palazzo Torlonia, 17 January 1875

Excellency!

I had just finished the introduction to my small publication "Der Ursprung der Wirbelthiere und das Princip des Functionswechsels" [The origin of vertebrates and the principle of shift of function],[80] a copy of which you will find enclosed, when I received your letter with the splendid photograph of the highly respected old man who neither looks nor seems to be ill-humored. Even though the eyes no longer take in light from the outside, within them there is still much more light than in many modern prophets who are equipped with natural and artificial lenses and who believe in nothing else but their own revelations.

But what will you say about my publication? Will you permit the introduction to be printed with your name and addressed to you, as I am sending it to you? This is my hope, the more so since I have expressly tried to characterize my opinions as being rather opposed to yours, so that you cannot be accused of having goaded me into such a revolutionary work. I wanted to consider myself close to you and not among the orthodox Darwinians only in the sense that I am far from admitting

---

[79] Von Baer says on the last page of his preface: "I would be satisfied if it would be considered as my contribution to have proved that the type of organization is responsible for development. Many others will win a prize. But the palm will be won by the fortunate man for whom it is reserved to trace the formative forces of the animal body back to the general forces that direct the life of the universe. The tree from which his cradle will be carved has not yet germinated" (translated from the German; von Baer, 1828, p. XXII).

[80] A. Dohrn, 1875. Dohrn addressed his introduction (pp. I–XV, Naples, January, 1875) "An Carl Ernst von Baer."

that the range of principles that the development of life urges us to seek is exhausted with natural selection; that therefore the metaphysical needs may not be denied their right as long as the physical explanations are so insufficient.

I understand, or believe I understand what you desire when you say that you cannot imagine development without purpose. I have tried to incorporate this element of striving toward a goal into the notion of perfectibility and in this way to render the teleological as well as the causal elements as linked evenly and uniformly to each other. At the moment I refrain from all effort to define this notion and I am also convinced that a satisfactory definition is not possible at the moment. But the need should at least be mentioned, and as a Darwinian without fear or reproach I felt I had the right to do so.

Therein and in reference to unlimited degeneration that at present is and has always been prevalent, I see the lasting elements of my work. Whether Amphioxus is an original or a degenerated fish, whether the ascidians should be understood in this way or another, this may be the major issue at the moment for the great mass of naturalists and they may tear one another's hair out about that–but if I should have succeeded in finding, with the definition of the shift of function, a truth that necessarily will produce other truths, then I shall happily be mistaken in many details.–

The printing of the little article, that will be published by Engelmann, will be completed in three weeks. I shall send you a galley proof. It contains three parts: 1. The genealogical connection of Annelids and Vertebrates. 2. The genealogical connection of Vertebrates and Ascidians. 3. The principle of shift of function. The last page contains a glimpse into infinity which will probably be very much resented,–but I have already been denied all qualification for natural science some time ago, because . . . I performed too much music! αὐτὸς ἔφα at Jena.–

That you are so kind as to remember the Zoological Station again in such an appreciative way is a double pleasure for me, since at the same time another recognition arrived, with the announcement that the Kaiser has signed on 2 January the decree which for the second time grants the Station 10,000 Taler from the Reich. This will pay a large part of our debts. The guaranteed income has also increased, I even hope soon to reach a perfect state of equilibrium between debit and credit.

But on all this I hope to be able to report to you personally next May, because I shall go via Vienna, Pest, and Moscow to Petersburg and shall not bypass Dorpat a second time. I shall try to rent out up to 20 or even more tables with fixed contracts–only then will my enterprise be safe.–

May I now ask you whether I receive permission to print the introduction I sent you? Of course I expect this permission only after you have read the text itself, which will be sent to you as galley proofs from Leipzig. At the same time I am sending the manuscript of the introduction to Engelmann, and I would ask you to send him a few lines: print, or do not print, the introduction.

Whatever you decide, it remains the truth that spiritually my essay has been addressed to you. Forgive my boldness.

And now I wish you, above all, the continuance of the freshness of mind and the serenity that speak from your letter, and I wish that soon, through the publisher, what you have to say about Darwinism may be accessible to us.

With heartfelt thanks repeated once more, and with the assurance of my immutable devotion,

Yours,
Anton Dohrn

---

[28] Dohrn to von Baer
6 February 1875, [Naples]

Excellency,

I hasten to make up, as far as I can, for an oversight of the publisher Engelmann and to write to you that I had asked Mr. Engelmann *not* to give the dedication to you to the printer, *before you had expressed your consent* and granted permission to print it.

I would now like to ask you with all my heart to let me know frankly even the slightest objection you may have against it,—because insofar as I might wish the little article to find your consent, so much I am deeply convinced that such liberties as exist in dedications and other purely personal matters should be handled only in the most considerate way. Good manners and decency are often so ill-treated nowadays that on no account do I wish to appear before your eyes as someone who would participate in such a mental attitude. I do not care about the opinion of the masses at all, but I do care deeply about yours.

You realize that this letter has been dictated by the selfishness not to be reproved by you,—and you will believe me that this is the honest expression of my feelings.

In great haste in order not to be too late,
Truly yours,
Anton Dohrn

Palazzo Torlonia, 6 February 1875

[29] Dohrn to von Baer
20 February 1875, Naples

Naples, Palazzo Torlonia, 20 February 1875

Excellency!

Although it will seem immodest, if I write to you again without first waiting for your answer, I have to shift the blame for that to 29 February 1792, on which day—according to Carus & Engelmann's *Bibliotheca Zoologica*—the author of the *Entwickelungsgeschichte der Thiere* was born.[81]

To express due respect to him on his 83rd birthday is the task of these few lines, the author of which after having done so disappears immediately again from the stage, although he will not omit to sign as

Yours truly and respectfully,
Anton Dohrn

[30] von Baer to Dohrn
8/20 February 1875, Dorpat

Dorpat, 8/20 February 1875

My dear Dr. Dohrn!

I have received not only your kind letter, but also, from Mr. Engelmann, the proof of the dedication as well as the first and third sheet of the text. I can only thank you for your letter to me because it honors me beyond merit. I have therefore gratefully given my consent. Of course I have also looked partly through the treatise. And there has been, as you assumed, no lack of head-shaking. That the Annelids turn themselves over to become vertebrates—this I was already somewhat familiar with, and found it therefore less objectionable. But that the extremities, and namely those of the vertebrates, have been derived from respiratory organs, this I find very objectionable. Are animals then supposed not to be able any longer to form anything from their own needs, but only to modulate what they inherited? If free motility is based on the nature of the animal, and if methods for obtaining this are provided in the water in many different ways, then it seems that for animals which are sup-

[81] At the beginning of his Autobiography von Baer himself comments on the difficulties with regard to the exact date of his birthday: according to the old, Julian, calendar (see note 5) von Baer was born on 17 February 1792, which according to the new, Gregorian, calendar would be 29 February, 1792 having been a leap-year. But since before 1800 the time difference was only 11 days, his birthday should therefore rightly be celebrated on 28 February (von Baer, 1866, p. 3).

posed to find their food on solid ground there have to be either wings or some other kind of support ["lever wheels"].

But I see well that the new way of viewing the world is completely different from the one to which I have become accustomed. Therefore I no longer wish to make myself heard in the Areopagus of Naturalists, if only before then I can turn loose an article on Darwinism. I do not know whether I wrote you that for quite some time I have, *dictando*, prepared such an article.

But now it is quiescent, and my publisher cannot make up his mind how to go about having adequate proofreading. But those are *curae minores* that should not pass beyond the borders of Dorpat. This treatise on Darwinism does not take a position for or against the latter, but balances objections and counterobjections against each other. I cannot help but find transmutation probable to a high degree; but I cannot declare Darwin's hypothesis of selection to be sufficient and have believed therefore that transmutation should be explained as a developmental phenomenon. Should I still live through the autumn of this year, I would be very pleased to make your personal acquaintance, I have even in the middle of this winter, at which time I am usually very fond of traveling, already planned a trip to Naples, i.e., not for the winter, but for the summer or autumn. But I am afraid I shall not be allowed to travel since my family always finds me even more fragile than I do. However, at the moment we are still surrounded by deep snow, what the future will bring can only be seen when the snow has gone, this year it seems to be very permanent.

With hearty respects,
Yours,
Dr. K. E. v. Baer

[31] Dohrn to von Baer
1 April 1875, Naples

Naples, 1 April 1875
Palazzo Torlonia

Excellency!

My little publication should be in your hands by now, may I therefore repeat my thanks to you for having granted me permission to print your name at its beginning.

I would have wished, however, that you would have found it possible to approve at least of the way in which I hope to attain genealogical results, and to approve the principle that I believe I have discovered. The results of my speculations certainly deviate very far from what is accepted today,–but is that in itself a mistake? Each day we live a new "today" and should I have paved the way for progress for today's today—how long will it merit this name, in the face of the immense regions where there are no signposts to guide us? I hope that "shift of function" might be, however, for the next few years a great help in locating and solving a series of morphological and physiological problems,—but it remains just one more concept of modality—and life, in the end, still remains incomprehensible to us.

But all this will not make my publication more sympathetic to you and I must console myself with the knowledge that in spite of your differing opinion I need not cease to see in you and in Darwin the two cornerstones of the present edifice of biology, and to hope that further development of our knowledge will also harmoniously smooth the differences that still exist between the mode of thinking laid out by you and the theory of natural selection.

As for myself, Excellency, you should always consider me as one of your greatest admirers who honestly tries to follow you in those ways of thinking that from the beginning seem to differ from his own basis,—and who wishes at least to understand you although he cannot agree with what he has understood.

And I wish therefore that you might soon search for a better environment for your life than that of Russia's snow and ice, and I volunteer for any escort service in case you really put your intention into practice, to pay a visit to us children of the South.

With assurance of my truest devotion,
Your Excellency's
Very truly,
Anton Dohrn

---

[32] von Baer to Dohrn
11/23 June 1875, Dorpat

Dorpat, June 11 respectively 23, 1874[5][82]

Honored Doctor Dohrn!

I regret with all my heart that my few remarks on your important publication seem to have disturbed you; I may also have been wrong in stressing too strongly my differing opinion of the origin of vertebrate extremities, and in having said too little about what I liked in your writing. Above all I find it much more attractive to derive the vertebrates from annelids, that are at least divided into segments, than from the com-

[82] The last number (4) has been underlined by someone else and corrected to "(5)". Heuss (1962, p. 223) quotes parts of this letter, though with mistakes and providing the wrong date (1 June).

pletely unsegmented sack-shaped ascidians whose muscle- and skin-layer form nothing but a sack. For this opinion I see no other reason but the observations of the development of Amphioxus and ascidians, [namely] that Amphioxus does not even seem to have developed [? *contrahirt*] from the ascidians. Derivation from the segmented worms[83] would strike the eye if an inversion of abdomen and dorsum could be proved. With the questionable breaking through of the mouth you have certainly struck at the Achilles' heel of this opinion; now all depends on whether such a breaking through can be proved or at least be shown as likely. I immediately remembered from the out-of-date stock of my memories Otto's *Siphonostoma diploehaites* [*sic*][84] which appears in the tenth volume of the Memoirs of the Leopold. Carol. Academy. I do not know at all what later became of the report on the mouth openings situated one behind the other, that is, how they can be sensibly interpreted. In Claus's handbook[85] I find no information about it. In other books, that I cannot examine now, there is nothing except referring them to other genera. As a resident of the Mediterranean coast and an expert on lower animals you will know this much better and I must assume that you expect no result from this line of thought since you do not enter into it.–

The principle of degeneration has also appealed to me very much, for if any change is possible then why not also a regression, which by the way we see so vividly in the parasitic crabs in the life cycle of a single individual. Nor do I have anything against shift of function, which to a lesser degree has been empirically demonstrated by Siebold in the air-breathing water snails.[86]–

–But there our views are different; I do not feel any need to derive the vertebrates from another larger group. Soft Amphioxus-like vertebrates may have existed for a long time without leaving behind distinct traces. I even suspect they are among some of the so-called *graptolites*. However, we differ completely in that you seem to assume, as do most Darwinists, that an animal can have no part that has not been inherited, that therefore our arms and legs must have been gill rudiments in the segmented worms, whereas I am convinced that every normal development of an organism (that is, with the exception of malformations and monsters) must be regulated to make existence on this earth possible. Should an organism therefore, according to my opinion, be constructed that has to seek for its food on *terra firma*, then linked to this is the fact

---

[83] Von Baer uses here the German word "Rundwurm" (literally: round worm) as a synonym for "annelids" (Ringelwurm) mentioned a few lines above, and is not referring to the separate phylum of *Nematodes*, colloquially called "round worms".

[84] Adolf Wilhelm Otto, "Siphonostoma diplochaitus," *Nova Acta Leopoldina* 10 (1821): 628–634. The different spelling in von Baer's letter could be attributed to his quoting from memory and/or his amanuensis' misunderstanding.

[85] Carl Claus, *Grundzüge der Zoologie*. Marburg, Leipzig: Elwert, 1868, 839 pp.; 2nd enlarged ed., 2 vols., Marburg, Leipzig, Elwert, 1872, 1171 pp.

[86] Carl Theodor von Siebold, "Ueber das Anpassungsvermögen der mit Lungen athmenden Süsswasser-Molluscen," *Sitzber. preuss. Ak. Wiss.* (1875): 39–54.

that it acquires jointed feet, and, what is more, according to the laws of mechanics. Now Darwin will say, if a grassfeeder or even a predacious land animal is born without legs, then a creature of this kind has to perish in the struggle for existence. I claim instead that such a structure is not possible other than through the strictest teratology.

–My article on Darwinism has indeed been sent off, but I cannot say that I am satisfied with it, because I cannot deny mutability, in favor of which I expressed myself even before Darwin.[87] However, I cannot declare myself at all in favor of Darwin's explanation of mutability, i.e., for the theory of selection, but I oppose it. It departs from the idea or the thought that in nature reason cannot be effective; whereas I cannot break away from my old view that nature reasons and strives toward goals, whether immanently or transcendentally. If someone wishes to have circles drawn as correctly as possible, and chooses the best ones from millions of trials that have been drawn freehand, these will still not be as perfect as those done with a compass, that is, when with a purposive necessity a curve is guided in an always equal distance around a single point. The second method will therefore achieve its goal better than selection, which has to reject an immeasurable number of attempts.–

–In a much earlier letter I have already expressed the hope of traveling south to Italy as a kind of leave-taking from the world. When I finally became liberated from my manuscript, I felt a free man. Unfortunately things seem to turn out differently. An ulcer has developed on my lower leg that had already been damaged in the past; it seems inclined to heal slowly or not at all. Before it has healed, however, of course I cannot travel. Should this happen in a month or two I could still start to travel across the border in autumn. I do wish to come to Naples but I scarcely dare to hope for it. In any case you should not let yourself be deterred by that, because I could still see your establishment and would find Mr. Kleinenberg, to whom I ask you to pay my respects. I am very much afraid that with my former information I have kept you from important enterprises. I therefore ask you urgently to disregard my most recent hope; it is also quite possible that I have only to make the big trip into the Land of the Shades. To see you here would of course be a pleasure to me, but no business can be carried out in St. Petersburg during these summer months.

–Be that as it may, let us not quarrel about Darwinism etc.; in any case

[87] In particular in: "Ueber Papuas und Alfuren. Ein Commentar zu den beiden ersten Abschnitten der Abhandlung Crania selecta ex thesauris anthrop. Acad. Imp. Petropolitanae.," *Mém. Acad. Sci. Petersb.*, sér. 10, 2, Sciences nat. 8 (1859): 269–346. There von Baer acknowledges possible evolution of organisms under the influence of their environment. He presented the treatise to the Academy on 8 April 1859 and communicated its contents also to Richard Owen and Thomas Henry Huxley during a trip to England (4 Aug.–28 Nov. 1859). He repeatedly stressed that it was only when he was in England at that time that he learned that Darwin was working on a major book on the theory of transmutation (von Baer, 1876, p. 248). On von Baer and Darwin, see Raikov, 1968, pp. 364–382, and many others (too many to list here).

by founding an establishment where development can be carefully studied because of the possibility of maintaining embryos alive, you have taken a step that will provide bridle and reins for Darwin's speculations if bridle and reins are needed. If you do not find the reins in Naples, then they are probably not needed at all.

With unchanged devotion and respect,
As ever, yours,
Dr. K. E. v. Baer

---

[33] Dohrn to von Baer
8 July 1875, Hökendorf

Hökendorf near Alt-Damm
Stettin, 8 July 1875

Excellency!

Nothing at my father's home could have been a greater and more pleasant surprise than the long and friendly letter that you sent me.

It is quite true that I regretted very much not having obtained grace with you through my hypothesis on the origin of vertebrates. I believed you would recognize with sympathy the great grasp that this hypothesis encompasses, looking forward and backward, and that you would see in the notion of perfectibility that I postulated a kind of reconciliation between striving for a goal, i.e., teleological necessity and causal necessity. And I frankly admit to you that I would like very much to be able to present to you orally the thoughts I withheld, not only the speculative but also the factual ones which in a second publication, in a morphological-genealogical explanation of our sense organs, beginning with eye, ear, and nose, will assure my hypotheses of such preponderance and power of explanation that it will calmly drive from the field all and every objection. But at the moment I am still the slave of the Zoological Station and have to perform drudgery, i.e., to travel back and forth in the wide world and provide, as guard against the vicissitudes of fortune in the future, the means to assure what has happily and successfully begun. Related to this is, first of all, the fact that my countrymen, in recognition of the national element involved in the enterprise, are giving me 10,000 Taler toward acquiring a small steamship and several pieces of equipment for the edifice of the Station, but then I have to increase the number of tables firmly subscribed for up to 24, in order to cover the running expenses (at least barely, as they amount to 13,000 Taler); the income of the aquarium can be used for paying debts and for setting up a reserve fund. I hope to have reached this goal in 1877, negotiating with Germany for 10 tables, Russia and Italy for 4 tables each, Austria-Hungary for 2, and the other European and extra-European countries, with alternative contracts, for the remaining tables.[88] Only then shall I shout *exegi*. Italy has already, through the official voice of its Minister of Education and through a document signed by all its universities, assured the support of the 4 tables for the coming year and, at the moment, will sign the respective contract for 5 years; the separate German governments will next year have a total of 10 tables and I hope my diplomatic skills will succeed in attributing all these 10 tables to the Reich, and will induce the separate states instead to donate, from their own funds, fellowships for the investigators who go to Naples. Since I have come into closer personal contact with Minister of Finance von Reutern, and since his personal interest in the Zoological Station has been aroused I hope it will not be difficult to obtain from him the money for supporting 4 tables, if Mr. von Bradke and Minister Tolstoi become convinced that Russia has more than enough biologists to keep this number of tables occupied.

Eight days ago I was in Vienna and Pest, I hope with the desired success; now the need for subscriptions drives me all over Germany, and worry over the need for more tables drives me to Copenhagen and Stockholm as well as to England.

There you have, Excellency, the whole battle plan. An open circular letter that I have addressed to Professor von Siebold[89] will introduce you to several interior scenes of the Station, which I hope, together with the oral reports of Dr. Rosenberg, will provide you with a picture of the whole.

I cannot, however, believe that you should keep your suitcases ready for the trip to the Land of the Shades. News reports have brought me totally different tidings, and I am convinced that you will continue your march into the nineties with a firm step. I am counting with certainty on being able to present you with the proofs, in mint condition in fact, of the descent of fishes from annelids, although at the moment I am busy with mining the noble metal that awaits being pressed into coin during the next two years. But the matrices are more than ready—I am no longer "sicklied o'er with the pale cast of thought." Until then people

---

[88] In 1884 Dohrn had almost reached this goal, although the distribution was somewhat different from what he had envisioned. The 23 worktables subscribed for by then had been paid for by the following countries: the German Reich and its States 10, Italy 4, Great Britain 2, Russia 2, Netherlands, Switzerland, the United States, Belgium and Hungary one each. The following year (1885) five Italian and two German tables were added.

[89] A. Dohrn, "Mittheilungen aus und über die zoologische Station von Neapel. Offenes Sendschreiben an Prof. Dr. C. Th. von Siebold." *Z. wiss. Zool.* 25 (1875): 457–480.

Carl Theodor von Siebold (1804–1885), professor of zoology and comparative anatomy in Munich (1853), belonged to the group of senior scientists and correspondents of Carl August Dohrn who put their name and influence at the disposal of C. A. Dohrn's son for his scientific enterprise. Dohrn chose to publish an open letter because he wanted once and for all to report publicly on the development and facilities of the Zoological Station, and at the same time to counter such rumors as those saying that he wasted government money, that the Institute was infested by rats, and that privacy for the investigators in the main laboratory and the supply service of marine organisms were highly unsatisfactory. Dohrn also stressed the importance of the Aquarium for the observation of the behavior and life habits of marine organisms, an opinion that von Siebold fully shared with Dohrn.

may call me, without fear of contradiction, visionary, uncritical, ignorant, and what else are the flatteries that dear colleagues are holding in readiness—I know all those nice epithets; they have also opened the avenue along which I proceeded toward the Zoological Station. It is not very important.

But with loyal and grateful respect, Excellency, I press your hand and hope to see you in Naples or Dorpat before we have grown one year older.

My father and my brother Heinrich both ask me to give you their warmest wishes and I ask you to be kind enough to remember me to scientific colleagues in Dorpat, in particular with kind regards to Dr. Rosenberg, whose presence here in the Zoological Station was a great pleasure to me, but also a great disappointment insofar as everything was not arranged far enough ahead of time to permit such an excellent worker the full success of his efforts.

I shall not fail to report on further progress in favor of the Neapolitan Station; I have also to report that Agassiz's creation, the Anderson School of Natural History, is going under, since they cannot procure any further public or private funds.[90] I hope this time Europe will show America how to do it.

With repeated hearty greetings, and thanks,

Truly yours,
Anton Dohrn

---

[34] Dohrn to von Baer
18 August 1875, Stettin

Stettin, 18 August 1875

Excellency!

This time the representative of the Zoological Station approaches you with a really enormous request.

I believe that I have already written you that I have for some time past been busy in obtaining further financial means in Germany through subscriptions.[91] Unfortunately the matter meets more obstacles than I like or than are advantageous to the Station. General commercial crises serve everywhere as justification, so that even very wealthy people promise either nothing or very small sums; and I consume disproportionately much strength and time to procure insufficient funds.

Just back from England, I have convinced myself that a small steamship, that would offer extraordinary advantages to the Station in many

---

[90] John Anderson suddenly discontinued his generous financial support of the Anderson School (A. Agassiz to Dohrn, 11 June 1875, Cambridge, *FA: Ba 662*; see note 42).

[91] See note 59.

ways, cannot be obtained for less than 800 Taler. On the other hand, I need 800 Taler to pay for the acquisition of a steam air pump, and also 800 Taler to buy new thick panes of glass for the aquarium. The organization of a collection (procuring cupboards, alcohol, glassware, etc.), a small amount of new construction in the basement to create a coal storage room, a small shop for my mechanic, and a store room for preserved animals, as well as the construction and equipping of six new worktables, amount to several thousand Taler,—in short all in all the 12–15,000 Taler that the German public would give me to use will just cover the needs of the young Institute begging for nourishment.

It is of course a pure matter of trust that such means are given to me. But after even the most distrustful and *ex officio* the most careful (and rightly so) factor, namely the State, has proved its trust with 2,000 Taler, and after the first year of activity has proven that such a trust was justified, it is, I believe, no longer to be expected that justified distrust should persist.

Would it be possible for you, Excellency, through your word and request in the circles of nobility and wealthy merchants of the Baltic provinces, to procure a substantial contribution to such a subscription?

This is one large request.

And now to the second. The number of worktables subscribed to absolutely has to be increased. I have equipped six new tables and must rent them in order to be able to cover, through the yields of their rental, all of the running costs. The aquarium yields 5,000 Taler; this amount is necessary for paying the last debts of 15,000 Taler and for creating a capital reserve of at least 5,000 Taler.

Twenty-four rented worktables yield 12,000 Taler; the regular running expenses amount to 13,000 Taler; therefore with *great* economy the income from the aquarium may fail once without immediately jeopardizing the existence of the Institute. But the twenty-four tables are indispensable, and this from next year on, i.e., beginning already in October 1876. I have explained this in Rome and in Berlin and I have found fully receptive ears in both places. The Italian Minister has requested a testimonial from all Italian universities, and on their unanimous vote he has put into the next budget four instead of two tables, and in Berlin I have been promised that next year an increase in the number of tables will be arranged there as well. Furthermore my contacts in St. Petersburg are very favorable; I have a friendly relationship and correspondence not only with Mr. von Bradke, but—which in this case is even more important—also with the Minister of Finance in person: i.e., I have exchanged several letters with Mr. von Reutern and I know that his Excellency takes a personal interest in the Zoological Station and has already been prepared for the fact that an extra request is going to reach him.

After some reflection it seems to me that you, Excellency, could carry through the doubling of the tables with a purely personal request, if you are inclined to take such a step. I myself cannot submit this request. To reconcile the many conflicting Russian scientists, societies, and universities would again be a very difficult, extremely laborious effort, and perhaps in the end nevertheless in vain, and whether such a joint declaration, if it ever came about, would have the emphasis that, with the persons in question, your single vote would have, remains very doubtful. In addition there are the formal protests of Russian scientists and universities against the stipulated closing of the Zoological Station from 20 June to 20 August of our calendar. I would be willing to cancel these holidays for Russia if this concession on my part would be responded to by an increase of the number of tables from 2 to 4—thus the Ministry would be humoring the wishes of their own nationals.

I report all this just *brevi manu* to you, Excellency, since I am bold enough to believe that for you I do not need any superficial decoration of my requests; you will know best whether what I ask lies in the reach of your power and also of your will.

I cannot hesitate. The fate of Agassiz's Zoological Station shows clearly that danger lies in delay—on the other hand I am so much convinced of the vitality of the Station and put such great hope in its scientific productive power, that I must summon up all my strength for the completion of its equipment.

Often my moves are also stimulated by the consideration that for the moment the young Institute cannot do without my personal abilities, even though I take all measures to make the Zoological Station capable of surviving my death should it be brought about by typhus, cholera, or some other accident. If it causes *me* great difficulties to carry the business through to the end, it would be almost impossible for someone else, unless he too could again place his own considerable means into the scales of success. And until now I see nobody who would be willing to do so,—only many of little faith, skeptics, critics, and whatever else you want to call the *gentes minores.*

I find it significant that I have actually found more enthusiastic and active support among those who belong to past times and who indulge past currents of mind than among my equals in age. But perhaps this may be explained in Goethe's soothing words:

> Lebst im Volke; sei gewohnt
> Keiner je des Andern schont.[92]

But then you see quite frequently that you yourself do not spare others, and you are not surprised to encounter some opposition.—

I plan to remain in Germany until the end of September and to contribute, as far as my strength permits, to the good success of the subscriptions. The Station is in very good hands and once all the financial and organizational difficulties are solved, the administration of the Institute in itself offers no significant difficulties at all.—

Hoping, Excellency, that this letter will find you again in completely untroubled physical well-being, and that it will find friendly acceptance,

---

[92] Motto chosen by Goethe for a group of poems entitled "Sprichwörtlich".

I greet you with all the heartfelt devotion that I have always felt and that I have tried to combine with the due respect that the young adept owes not only to the unattainable master, but also offers out of his own innermost need.

Should the summer or autumn not take me to Dorpat's degrees of latitude, I count with fair certainty on next spring undertaking a trip through Russia via Constantinople, Odessa, and extending it to Petersburg and Dorpat.

My father sends his kindest greetings and hopes for good news from you.

With the same hope I close this letter
as Your Excellency's
most truly devoted
Anton Dohrn

---

[35] von Baer to Dohrn
16/28 August 1875, Dorpat

Dorpat, 16/28 August 1875

My dear Doctor Dohrn!

Although I do not know where and when I can find you, I nevertheless hasten to answer your dear and honored letter rather quickly. I will address this letter to Stettin hoping that it will be forwarded to you from there.

Unfortunately what I have to say is not very comforting. I consider any significant subscription from here completely unfeasible. Of the nobility the far greater part is reduced to poverty. There are a few wealthy and even very wealthy men among them; however, all members of the nobility have now to meet a growing public demand for the establishment of elementary schools, and they make considerable contributions in this area. One corporation, however, has recently earned significant funds, that is the body of merchants of Reval, for which the Baltic Railway has opened an extensive sphere of activity. But I have never heard that the Reval body of merchants has ever shown any interest in any scientific establishment. Here in Dorpat there exists a Society for the Promotion of the Natural History of the Baltic Provinces, of which I have the honor to be the very useless president. This society has been laboring for years, close to extinction because of lack of funds. But never has a merchant from Reval or from wealthy Riga made a contribution to it. It scrapes a living from contributions by professors and schoolteachers.

But the other way of doubling the subscription for Russian naturalists is also barred to me. I had asked the Minister of Education for a subvention for the local Natural History Society, and received a negative answer although this subvention had been granted to all other Russian universities for founding Societies of Natural History. Your own steps are more likely to be successful. If you knew Russia sufficiently, you would also know that such a request will be granted more easily to a foreigner than to a Russian countryman. One wishes to be praised abroad. This is the reason for the Sinai Bibles[93] and similar enterprises. Moreover I must tell you that Dr. Rosenberg has written that he did not find ready the table assigned to him but received it only later. He did not write this to me, although he had received his acceptance via my correspondence with Mr. von Bradke. Therefore I also was informed right away about his complaint; I do not know whether this has also reached von Bradke. For support of your request in Petersburg could you not turn to those Russians who have visited your Institute and who have worked there? Such persons would be able to describe most emphatically the advantages that a stranger has when affiliated with you, rather than arriving as an independent naturalist.

With the heartfelt wish that you will see in this "*non possumus*" not indifference but only some insight into local conditions, and with the plea to keep me in friendly remembrance,

Your
Respectful
Dr. K. E. v. Baer

---

[36] Dohrn/CAD to von Baer
3/4 September 1875, Stettin

Stettin, 3 September 1875

Your friendly letter, your Excellence, awaited me here yesterday, on my return to my family from a short journey to Copenhagen and Stockholm. I went to Scandinavia in order to provide for my enterprise there as everywhere, and I hope that the local naturalists will succeed in winning support for me.[94]

I understand all too well what you write about the impossibility of doing something in your country. Also it does not discourage me, I am used to confronting such difficulties, especially in this financially difficult period. Only undaunted persistence can win my final victory for me,

---

93 Konstantin von Tischendorf (1815–1874), professor of theology at Leipzig, was mainly working on a text reform of the New Testament. On several trips in Europe and in the Middle East he collected material for his research; in 1844 and 1859 he discovered the *Codex sinaiticus* (end of the fourth/beginning of the fifth century). A first edition in four volumes was published in 1862 in St. Petersburg; numerous further editions appeared in Leipzig.

94 Dohrn was too optimistic; only in 1922 was the first Swedish worktable subscribed to (Partsch, 1980, pp. 244–257).

and I do not have to affirm that I am firmly decided to assure this victory *à tout prix.*

I cannot prevent small vexations from presenting themselves at every step, and therefore I also cannot hinder Dr. Rosenberg, for example, from saying and writing things that, more loyally and truthfully, he should have either omitted completely, or presented along with all the explanatory factors. Let it suffice, though, Excellency, to tell you that Mr. von Bradke is far from putting the blame for this inconvenience on me since he knows perfectly well that it was the Ministry's fault. In a very cordial, almost friendly letter of 2/14 July he writes to me the following on this subject: "– –I hope that similar time conflicts in the use of the table will not happen again. Those that have happened may be attributed to the change of personnel in my department that I have had to effect.–" This was the answer to a remark on my part. Should Dr. Rosenberg take notice of this he would only do what is honorable.–

I am quite convinced that it would perhaps be wisest if I again conducted my affairs at St. Petersburg personally, and I plan to travel there at the end of October. On that occasion I shall also have the pleasure of paying you a visit in Dorpat,–and this is such an old wish that I put its fulfillment only after the urgent duties, that often force me to push everything else until later, in order to overcome first and above all the financial difficulties of the Zoological Station. You may then perhaps be so kind, Excellency, as to give me a few more good tips, and I can explain to you the situation of the whole affair better than through ten letters. I was pleased to deduce from the silence in your last letter that the wound on your leg, about which your second-before-last letter complained, seems to have closed, and that you have now regained your old splendid health. May it stay with you for a long, long time!

With this most heartfelt wish I am,

Yours respectfully,
Anton Dohrn

[. . .][95] I shall remain permanently in Germany until I have accomplished everything. The Station is in good hands.

[CAD:]
Excellens Ursa maxima
the verminous world won't lose anything if I interrupt a learned article on Australian paussids (and therein a bittersweet lamento on friend Westwood's boners in his monograph)[96] in order to write to you by hand, following the illustrious *Gaudissart de la Station Zodiacale,* a hearty greeting and handshake. Yesterday I had an attack of genuine English

---

[95] Seems to be "WR" [?] in the German original.

[96] Paussids are a small group of beetles inhabiting nests of ants. John Obadiah Westwood (1806–1893) was professor of zoology at Oxford and co-founder of the Entomological Society. C. A. Dohrn, "Ueber australische Paussiden," *J. Mus. Godeffr.* 12 (1876): 48–55.

high aristocracy, namely a slight podagra, today it is gone, but tomorrow there will be instead my wife's birthday on which occasion I want to give her a fur coat, without adding into it for her the appropriate louse—nothing is perfect in this lousy world except the respect with which you are being greeted by

Your always unchanged
C. A. Dohrn

4 September 1875

---

[37] von Baer to Dohrn
17 September 1875, Dorpat

Dorpat, 17 September 1875

To Dr. Dohrn jr.

My dear Doctor Dohrn!

Notwithstanding the hope of seeing you here, which would be a bright point in my solitude, I feel the need to write you beforehand so as to cancel an unfounded suspicion that I might easily have caused. I shall immediately come to the point, since introductions would bring even more importance and darkness into the matter. I had written that Dr. Rosenberg announced here that he had not found any worktable prepared for him. Dr. Rosenberg had not yet returned here. He arrived a few days later, and reported to me that through the fault of the Ministry you had received no notification of its disposition, as you also told me in your last honored letter.

Also, Dr. Rosenberg had not complained to me that he had not found everything prepared to receive him. The communication he sent here had reached my ear only indirectly, and I had indeed assumed that he had complained about it. His complaint could however only be directed to the Russian Ministry. You should therefore not record his objections permanently in your heart.

The ulcer on my foot has still not healed completely, but is so close to healing that there seems to be only a remnant remaining, the size of a pinhead. Unfortunately I cannot see that far.

I have gratefully received your report on the progress of the Stazione in the *Zeitschrift für wissenschaftliche Zoologie.*

With a hearty greeting until we meet in the future

Yours respectfully,
Dr. K. E. v. Baer

[38] Dohrn to von Baer
15 October 1875, Berlin

Berlin, 15 October 1875

Excellence!

Things look bad for my visit to you. I work like a horse,—no, like a steam engine,—on the acquisition of means for the Zoological Station; I also hope for success—but only through the most reckless perseverance and persistence. Thus my departure is postponed day by day. I have already given up Moscow, since I have to give a lecture on 6 November at the Geographical Society which is supposed to promote subscriptions throughout all of Germany. But I shall go to Petersburg *à tout prix* for support of two more tables. Then the business would be finished for quite some time and funds would be at my disposal with which I can thoroughly develop the whole Institute that has had complete success in circles of the German government and science. I suppose that depends mostly on Minister of Finance von Reutern and on Mr. von Bradke. With both gentlemen I am in personal, confidential relationship so that I can turn myself loose immediately in Petersburg. But, Excellency, do not be angry with me if in spite of your last letter I still insist that your intervention might further pave the way for me. I *know* that with both the persons just mentioned, your personal influence is far greater than that of the Academy and all the native Russian investigators. Should it therefore not be too much trouble for you to write to Mr. von Bradke, as well as to Mr. von Reutern, I am convinced that you would certainly facilitate success with the simple request to accept, if feasible, my request, which will be personally presented before long. I would of course bind myself to keep silent everywhere—which I have to do pretty often in order to maintain my influence.

Then my stay in Petersburg may perhaps also be shortened in such a way that on my way back I could pass through Dorpat, which I have longed to do so sincerely and for so long.

The reservations for this winter at the Zoological Station are again numerous, among others those of Carpenter, Pringsheim, Victor Hensen from Kiel etc.[97] Switzerland will probably also join us now.[98]

---

[97] William Benjamin Carpenter (1813–1885), British physiologist at the Royal Institution and London University. Carpenter came to Naples in January 1876 for two months, working at the Cambridge table.

Nathanael Pringsheim (1823–1894), plant physiologist and professor of botany in Berlin (1868), never worked at the Zoological Station; however, he visited it with his family in April 1878 (Groeben, 1985a, p. 128).

Victor Hensen (1835–1924), professor of physiology at Kiel (1871–1911) and a founder of quantitative marine biology, coined the term "plankton." Hensen worked at Naples from 7 March to 26 April 1876 at a Prussian table (Hensen, 1876).

[98] Switzerland subscribed to a permanent worktable only in 1878.

My address in St. Petersburg will be c/o "Alexander von Baranowski. Fantanka dom Fedoroff," he is my wife's uncle.

The acceptance of my publication on the shift of function has been far more favorable than I expected, here in Berlin as in Leipzig and above all also in England. There, there is even the *intention* to translate it,—whether this will come about, I have no idea. I most anxiously wish I could put aside the practical work to take up the theoretical, to prove the hypothesis I expressed in the theoretical realm. I hope I will succeed in doing so by the end of this winter. I shall probably need a couple of years, but then there will see the light of the day facts that will surely cause many mouths to fall open. Q. d. b. v.!

Goodbye, Excellence, hopefully I shall still see you and may be able to shake your hand gratefully.

Yours respectfully,
Anton Dohrn

---

[39] Dohrn to von Baer
15 October 1875, Berlin

Berlin, 15 October 1875

Excellence!

In today's letter I forgot to tell you that the Prussian government as well as the Berlin Academy will grant me funds for the publication of the posthumous writings of Casp. Friedr. Wolff—although for the moment *sub rosa!*[99]

I shall therefore pay a visit to Count Lütke[100] and submit my project to him with your approval, in order to make possible a provisional agreement. I would then undertake an edition of the complete works of Wolff's writings, since much has completely disappeared. I would be very glad to have your advice and opinion on that.

With heartfelt greetings and faithful devotion

Yours,
Anton Dohrn

---

99 There is no written evidence for such promises; they were probably made during personal encounters while Dohrn was in Berlin.

100 Friedrich Benjamin Count Lütke (1797–1882), Russian admiral, president of the Academy of Sciences (1864–1882). See Introduction, pp. 17–18 and related notes.

[40] Dohrn to von Baer 22 October 1875, St. Petersburg

Fantanka 13. Dom Fedoroff.
St. Petersburg, 22 October 1875

Excellence!

From one day to the next, I postponed my departure from Berlin until the evening of the 19th, when I came directly here. At the moment I do not know how I can finish things up here, because on November 4th I have to be back in Berlin for a conference on November 6th at the Geographical Society.

I hope you will not scold me too much for proceeding recklessly and straight toward my goal—it often costs me the greatest personal sacrifice. I have been wishing to see you and to talk to you for so long that I almost believe fate, along with so much success, wants me also to experience a deep felt failure by preventing me for the second time from meeting you.

And the reason that I have been kept for so long in Berlin is the difficulty of the *tot capita tot sensus.* Since I had to confer again with the Academy, I had to visit the members of the mathematical-physical class one after another, some of them even more than once, before I could say to myself that I had done enough to be able to count, to some degree, on success. Then there was still the Ministry of Education, and although things usually go much faster there, and decisions there are certainly helpful and in my favor, several things became delayed because of the absence of one important personality. This was about nothing less than the long-desired small steamer for the Station, and there seems to be reasonable hope that Berlin will provide the money for it.

I do not know how things will go for me here in Petersburg. Mr. von Bradke, whom I had tried to ask today for a conversation, is not *en ville,* he will be back only on Sunday. He will report to me the state of affairs. From your colleague Brandt,[101] I learned that indeed the Academy has not been further informed on the affair [i.e., worktables] nor did Prof. Brandt know whether Russian zoologists had worked at all at the Station. From this it may probably be deduced that the further treatment of the affair lies exclusively in the hands of the Ministry and that in the last instance Mr. von Reutern has probably the last word. This will probably soon be demonstrated, and I shall not fail to report on this to you, Excellency. Until then I take a leave of absence, and add only the request that you will be so kind as to forward to my address here letters that may have arrived for me at your place.

At the same time I hope to have a few words confirming your con-

[101] Johann Friedrich Brandt (1802–1879), German zoologist, professor at the Medico-Surgical Academy in St. Petersburg (1831), member of the Academy of Sciences there (1833).

tinuing health, and since one should never cease hoping, I also hope that what 1874 and '75 have denied to me, 1876 will grant me, and that I can finally see whether the bust that has been erected in the hall of the Zoological Station actually resembles you.

With heartfelt greetings,
Your very respectful
Anton Dohrn

---

[41] Dohrn to von Baer
2 May 1876, Naples

Naples, Stazione Zoologica
2 May 1876

Excellence!

The most extreme activity that has again overtaken me this spring has until now kept me from thanking you for sending me your book.[102]

I need not repeat the feelings that accompanied my reading your expression of opinions. The historical concept that emerges so strongly from your words now diminishes; because of prehistory, history is now all too easily forgotten,—as is the fact that one's self, with its thoughts and wishes, is a historical particle, the presence of which is as little permanent as anything else. Thus when suddenly a live voice comes and speaks, as for an eyewitness, of times gone by that are as remote from us as tertiary rocks that only through a few petrified ideas give us proof of a formerly existing pulsating intellectual life, then we might become aware that we are in relation to the future what they are to us—and we learn to understand better that we too will sink into the bottomless abyss with all the dreamed-up properties of insights and truths which still have to give proof that they can survive an intellectual epoch.

I still have to thank you personally for having in your postscript[103] subjected the ideas that have germinated in my head to such an indulgent judgment. It is true that Prof. Semper has resented this,[104] but anxiety regarding priority is probably also a sign of unhistorical culture, if not of a worse lack of culture. In these days it seems almost inconceivable that opinions may persist for years without having been rendered

---

[102] K. E. von Baer, *Studien aus dem Gebiete der Naturwissenschaften*. St. Petersburg: H. Schmitzdorff, 1876.

[103] K. E. von Baer, *Studien* (see note 102), Nachträglicher Zusatz, pp. 474–480; on Dohrn: pp. 476–479.

[104] Carl Semper, "Die Identität im Typus der Gliederwürmer und Wirbelthiere. Eine vorläufige Mittheilung." *Verh. phys. med. Ges. Würzburg* 3 (1876–1877): 102–112. In an extensive note (note 1, pp. 102–103) Semper documents that it was his contribution and not Dohrn's, as von Baer had said (see note 103), who had for the first time turned to the Annelids for investigating the relationship between invertebrates and vertebrates.

shallow and stale by half a dozen preliminary and as many "postliminary" [*sic*] communications. What a difference from those times when your publisher accused you publicly when he could not induce you to finish the second volume of the *Embryology*![105] Preliminary communications are like new weapons that can be fired seven times in one minute – this is also why one aims less carefully.–

Of the Zoological Station, I can report only good news. After it was greeted in the beginning with probably exaggerated expectations, a reaction [followed?], and above all in Germany all sorts of malicious gossip grew exuberantly which mostly originated out of personal malevolence. I could do nothing else except proceed unswervingly to complete the Institute. This winter has now brought here investigators of more mature judgment, and I have had the pleasure that Professors Hensen from Kiel, His from Leipzig,[106] Grenacher from Rostock,[107] old Professor Carpenter from London, further Reinke,[108] the botanist from Göttingen, and several others, have assured me that they have all been highly satisfied and, in private with governments and in public in journals, oppose the sneaky gossip with their favorable judgment. This should once and for all take care of the attempts to trip me up.

I have now published the first annual report, that mainly renders account of the administration of the Institute.[109] Soon the Institute will

---

[105] The first part of von Baer's classical work had been published in 1828: *Ueber Entwickelungsgeschichte der Thiere. Beobachtung und Reflexion. Erster Theil.* Mit drei colorierten Kupfertafeln. Königsberg: Gebr. Bornträger, 1828. XXII, 271 pp. The second part appeared only nine years later (Königsberg: Gebr. Bornträger, 1837. 315 pp. with 4 copper plates). On an introductory page of the 1837 volume ("Nachricht"), the publishers justified the 1837 publication by explaining that printing of the second part had already been started in 1829 and had proceeded up to folio 38 in 1834, continuing: "Until today we have not been able to obtain a treatise with which the author, who at present lives in St. Petersburg, wanted to close the volume, nor the preface and the figure legends, nor have our letters been answered during the last 15 months; we have therefore given up hope of receiving the last part of the work, and we feel the obligation to deliver this second volume, at the request of our clients, as it is. Königsberg, August 2, 1837. The publishers" (translated from the German). Von Baer himself explained the delay by the difficulties of his moving from Königsberg to St. Petersburg at the end of 1834; only during the following winter did he again have access to the relevant publications in his library, without which he could not finish his studies (von Baer, 1866, pp. 398–399; 448–451).

[106] In March 1876, Wilhelm His (1831–1904), professor of anatomy at Leipzig University, worked for one month in Naples at the Saxon table; he published his results in: "Ueber die Bildung von Haifisch-Embryonen," *Z. Anat. Entwickl. Gesch.* 2 (1877): 108–124.

[107] In December 1875 Hugo Grenacher (1843–1923), professor of zoology at Rostock, arrived in Naples; he stayed for five months, occupying the Mecklenburg table and working on the arthropod eye. ("Untersuchungen über das Arthropoden-Auge," *Klin. Mbl. Augenheilk.* 15, (1877): 4–42).

[108] Johannes Reinke (1849–1931), professor of botany at Göttingen (1873) and Kiel, worked from October 1875 to March 1876 as the first botanist at Naples. He published his results in:"Ueber das Wachsthum und die Fortpflanzung von *Zanardia collaris*, Crouan (*Z.prototypus*, Nardo)," *Monatsber. preuss. Ak. Wiss.* (1876): 565–578.

[109] A. Dohrn, 1876. The report contains not only a detailed description of the building of the Zoological Station, and of its construction, but also a sample contract for the use of a table, lists of laboratory equipment, lists of the first guest investigators, data concerning the sale of collections of preserved marine organisms and the collecting periods for marine animals.

be considerably enriched by the possession of a small steamship, for the acquisition of which the Berlin Academy and the Minister of Education have granted 8,000 Taler. The internal organization of the establishment has also made progress, the personnel are trained, routine replaces personal initiative, and slowly it becomes possible to suppress the purely bureaucratic to a level below scientific activity. Many important and influential men have visited the whole Institute this spring, it grows in popularity which gives me hope that I will more easily succeed in acquiring the still necessary funds.

Unfortunately no Russian has been here this spring. This eliminates the possibility of my asking the Russian government for an increase in the number of tables which the Minister of Finance von Reutern has already promised me. This is a great pity because the Institute needs an income of 15–16,000 [Taler] and has to try to procure this sum in large part from the subscription fees for the tables. I have no idea what could be the reason for the Russians suddenly staying away.

Allow me to send some ground plans of the Station with this letter.[110] Even if you, Excellency, should not be able to see them, and to recognize from them the organization of the Institute, I still hope that one of your relatives might give you a picture of what is going on after having read the Annual Report, and, as usual, it will as always fill me with pride to know that you continue to maintain for the Institute your interest and sympathy. The next Annual Report will bring more scientific fruits that are ripening in the Institute itself.

With unchanging respect,
Yours,
Anton Dohrn

[110] Dohrn enclosed in his letter five printed ground plans of the finished building (basement, first and second floor, two transverse sections); they were provided with detailed legends (in German) and signed by the Station's engineer Giacomo Profumo (d. 1898).

APPENDIX

Karl Ernst von Baer–Anton Dohrn Briefwechsel

[1] Carl August Dohrn an von Baer
10.9.1862, Stettin

Herrn Akademiker von Baer
Excellenz.

Hochverehrter Freund

Ihren geschätzten Brief wegen des an der Ostküste Rügens zu schöpfenden Seewassers fand ich vor, als ich im Juni von einer Reise nach England u. Frankreich heimkehrte. Ich gab sogleich nach mehreren Seiten Auftrag dieserhalb, weil ich durch mancherlei Arbeiten verhindert war, selber die Reise nach Rügen zu machen. Aber die Wahrheit des Sprichworts Aide Toi kam auch hier zur Geltung, die Beauftragten kehrten wieder *ohne* das gewünschte Salzwasser und es blieb mir nur übrig, den Besuch eines naturforschenden Freundes aus Paris zu benutzen, mit ihm Anfang Septembers auf entomologische Jagd nach Stubbenkammer zu gehen und bei dieser Gelegenheit auch die beifolgenden 2 Flaschen zu füllen.

Sie sind nun zwar nicht, wie Sie es wünschten, bei Arcona genommen, doch aber nicht weit davon, und da in der Nähe von Stubbenkammer kein süßes Wasser (mit Ausnahme höchst unbedeutender Waldbäche) in die Ostsee fließt, so wird der Salzgehalt mit dem bei Arcona vorhandenen Seewasser wohl so ziemlich identisch sein.

In wenigen Tagen begebe ich mich mit zweien meiner Söhne, wobei auch Heinrich, welcher die Ehre hat, Ihnen bekannt zu sein, zur Naturforscherversammlung nach Carlsbad. Es ist die Möglichkeit, daß wir Sie dort treffen, was uns sehr freuen würde. Aus diesem Grunde addressire ich diesen Brief an die Kais. Akademie, damit die Flaschen nicht die Reise vergeblich machen.

Dieselben werden am nächsten Sonnabend (13. Septbr n.St.[a]) von hier mit dem Dampfer *Trave* Capt. *Heidtmann* nach Petersburg abgehen und es wird daher die K. Akademie die Güte haben, sich bei Ankunft der Trave (den 16 Septb n.St.) bei der Anlegestelle in Petersburg die

---

[a] = neuer Styl.

Flaschen, welche in einer an die *Akad. d. Wissensch.* addressirten Holzkiste sind, abholen zu lassen.

Ihrem Wunsche, über die an der Rügenschen Ostküste vorkommenden zoologischen Producte Auskunft zu erhalten, wird Heinrich zu entsprechen suchen, soweit es ihm möglich ist. Dergleichen hat seine sonderbaren Schwierigkeiten, da man von den Profanen meist nur unzureichende Auskunft erhält, und ihnen fast alles speciell abfragen muß, wobei dann allerlei Irrthümer mit einfließen können.

Mich Ihrem freundlichen Andenken angelegentlich empfehlend und mit der Bitte, Ihrer Frau Gemahlin meinen Respect zu vermelden

Ihr herzlich ergebner
Dr. C. A. Dohrn.

Stettin 10 September
1862.

---

[2] Heinrich Dohrn an von Baer
3.6.1863, Stettin

Stettin 3. Juni 1863

Hochgeehrter Herr und Gönner

Schon während des Winters hätte ich Ihnen Bericht abstatten sollen über das, was ich bisher in der Sache über Ostseefauna zu Wege gebracht habe, aber da ich gern meine gewonnenen Resultate selbst eingeschickt hätte, so habe ich bisher noch immer gezögert. Dazu ist nun noch ein Augenübel gekommen, das an sich zwar unerheblich war, aber die bekannte Dauerhaftigkeit der Krankheiten dieses Organs besass, so dass ich wochenlang von jeder das Auge anstrengenden Thätigkeit absehn musste, natürlich zunächst von Lesen und Schreiben. Jetzt, Gott sei Dank, bin ich wieder hergestellt, und werde nach Erledigung einer entomologischen Sache zunächst mein Theil Ostseefauna fertig zu machen suchen; ich fürchte nur, dass es überaus lückenhaft bleiben wird; wenigstens für die niederen Thiere mit Ausnahme der Mollusken ist das Material mehr als dürftig.

Noch eine Frage, die Sie an mich gerichtet, habe ich zu beantworten – wegen meiner Conchyliensammlung. Dieselbe fängt allerdings an, zu den grossen Privatsammlungen zu gehören, da sie ungefähr 10,000 Arten umfasst; ich bin deshalb auch wohl in der Lage, Ihnen über die kleinen Sachen aus dem Asow'schen Meere so weit als möglich Auskunft zu geben, will aber gleich hinzufügen, dass die dortige Fauna sehr wenig bekannt ist, und deshalb höchst wahrscheinlich manches Unbeschriebene enthalten wird. Ist doch der Griechische Archipel und die Kleinasiatische Westküste fast eine Terra incognita an Conchylien.

Wenn Sie mir daher die mitgebrachten Sachen auf einige Wochen zur

Untersuchung anvertrauen wollen, so bin ich mit Vergnügen bereit, die dortige Fauna nach Ihren Wünschen etwas durchzuarbeiten.

Vater lässt bestens grüssen, und rechnet fest darauf, Sie im September zur Naturforscherversammlung hier zu sehn, um unsre Versammlung besonders zu zieren; wenn Sie nicht kämen, so würde er ein Spottgedicht auf Sie machen; ich meinerseits hoffe, dass Sie sich einer so bedenklichen Eventualität nicht aussetzen werde[n], und kann nur sagen, dass ich mich unendlich freuen würde, Sie hier zu sehen.

In aufrichtiger Hochachtung
Ihr ganz ergebner
Heinrich Dohrn

[3] Carl August Dohrn an von Baer
6.11.1864/12.2.1865, Stettin

Dem Herrn Akademiker
Dr jubilatus von Baer
Ehren Mitglied des Stettiner Entomologischen Vereins.

Excellenz!

Die Pommern müssen immer was Apartes haben! Wo Andre mit den Bajonneten [*sic*] sticheln, flegeln sie mit den Kolben, und wo Andre am 9 September ängstlich Tag und Stunde halten, gratuliren sie zum Doctor Jubiläum in aller Gemüthlichkeit erst am 6 November. *Honni [sic] soit, qui mal y pense!* Warum mußte auch die Gießener Naturforscher Versammlung gerade in den September fallen, und warum wurde auch der amtliche Bericht der Stettiner Versammlung erst jetzt fertig?

Indem wir ein Exemplar des letztern unserm gefeierten Ehrenmitgliede hiemit ehrerbietigst überreichen, knüpfen wir daran den aufrichtigen Wunsch, daß unserm Vereine die *Ursa Major* noch lange leuchte in körperlichem Wohlbehagen und mit der unverringerten geistigen Frische, welche unsern Heros in der Wissenschaft vor vielen Andern auszeichnet. Darauf ist heute bei der Feier unsers Stiftungsfestes ein schäumendes Glas von ganzem Herzen geleert worden!

Stettin 6 November
1864.

Im Namen und Auftrage des
Stettiner Entomologischen Vereins
der Praesident
Dr. C. A. Dohrn

Dr. Heinrich Dohrn, welcher morgen eine naturwissenschaftliche Reise nach den Cap Verde-Inseln antritt, schließt seine herzlich ergebnen Grüße den meinigen an.

Vertas quaeso.

Stettin 12 Februar 1865

daß auch das vorstehende Retardat wieder in den Abgrund der Verschleppung gefallen ist, ist weniger *meine* Schuld als die der blokirenden Dänen und die getäuschte Erwartung, daß Jemand diesen Brief und die dazu gehörige Denkschrift auf dem Landwege in Ihre verehrten Hände bringen wollte, dann aber nicht abgeholt hat. Aber nach dem weisen Worte *mieux vaut tard que jamais* schicke ich Ihnen heute wenigstens den Brief, und wenn ich die bevorstehende Reise nach Paris u[.] Palermo glücklich gemacht und *Diis faventibus* im Mai wieder mich *ad penates* zurückverfügt habe, so wird der T. nicht wieder hinderliche Eier in die Dampfschiffahrt legen und das Ihnen bestimmte Exemplar des Naturforscherberichts mit neuem Sequester chicaniren.

Jedenfalls kann ich Ihnen nun heute mittheilen, daß Dr. Heinrich auf den Cap Verde Inseln *feliciter* angekommen und daß er nach einem Aufenthalt von etwa 4 Wochen schon einen 8 Bogen langen Brief losgelassen, laut dessen er mit seinen Erlebnissen und Ergebnissen ganz zufrieden ist, allerdings mehr mit dem malakozoologischen Theil als mit dem entomologischen; doch das war *a priori* vorherzusehen. Die tropische Hitze behagt ihm ganz wohl und er leistet Bedeutendes im Vertilgen von Orangen.

Bewahren Sie, hochverehrter Freund, ihm und mir Ihr freundliches Wohlwollen und seien Sie herzlichst gegrüßt

von Ihrem aufrichtig ergebnen
Dr. C. A. Dohrn

---

[4] Anton Dohrn an von Baer
30.12.1869, Stettin

Stettin. 30. Decemb. 1869.

Hochverehrter Herr!

Glücklicherweise bin ich Ihnen nicht ganz unbekannt,—so dass ich der Nothwendigkeit überhoben bin, erst meine Legitimationskarte vorzuzeigen.

Mich treibt zu diesem Briefe ein Plan, der Ihnen bereits bekannt geworden ist durch meinen Freund Miclucho-Maclay. Ich beabsichtige, zoologische Stationen zu gründen, welche den an das Meer gehenden Zoologen alles Handwerkszeug, Boot, Schleppnetz, Taue, ferner Aquarien und schliesslich sogar ein Haus mit Wohn- und Arbeitszimmer bieten, wofür nur eine geringe Summe von jedem Benutzenden zu zahlen sein soll.

Ich gehe im nächsten Herbst wieder nach Messina, um mich mit der Entwicklungsgeschichte der Anneliden eingehend zu beschäftigen. Bis dahin möchte ich so viel Geld gesammelt haben, um zu gleicher Zeit den Bau der ersten Station unternehmen zu können. Ich schätze die Kosten vorläufig auf 2000 Thaler, die ich durch Sammlungen aufzubringen gedenke.

Ich brauche Ihnen nicht zu sagen, hochverehrter Herr, welche Hilfe es für mich sein würde, wenn Sie mir in einem Briefe, Ihre Theilnahme und Billigung meines Unternehmens aussprächen. Ihr Name würde mich überall legitimiren und ich dürfte hoffen, bei Leuten günstige Aufnahme zu finden, die mich nicht kennen und nicht beurtheilen könnten, ob ich Etwas Nützliches unternehme.

Da ich durch meine Studien an Krebsen und an andern Seethieren hinreichende Erfahrungen habe sammeln können, was für ein zoologisches Laboratorium nothwendig angeschafft und eingerichtet werden muss, so denke ich, dass Sie dreist es wagen dürfen, mir ein Vertrauensvotum auszusprechen. Durch viele Verbindungen mit Fachgenossen denke ich dann später es auch dahin zu bringen, dass eine zoologische Bibliothek auf der Station eingerichtet wird, die ja für die reisenden Zoologen von grosser Bedeutung ist.

Ich ermüde Sie nicht mit Aufzählung alles Dessen, das sich einrichten lassen könnte: Sie wissen es besser als ich, und es steht mir mehr an, Ihrem freundlichen Rathe zu folgen, als Vorschläge zu machen.

Darf ich Sie also noch einmal, um diesen Rath bitten? Sie würden ihn keinem Menschen geben, der in die Goethe'sche Kategorie der "Schiefohren" gehört,—und dankbar würde ich durch und durch sein.

Ich sende zugleich ein erstes Heft meiner Arbeit, das mit keinerlei Anspruch bei Ihnen auftritt, sondern nur den Verfasser als thätig in den Wegen der Entwicklungsgeschichte bei dem Altmeister derselben anmelden soll. Ich hoffe im Laufe des Sommers 2–3 andre folgen zu lassen.

Verzeihen Sie diesen Raubanfall auf Ihre Zeit einem jungen Manne, der die Freude an seinem Berufe mit der Verehrung für Ihre Person innig verbindet.

In völliger Ergebenheit
Ihr
Anton Dohrn
der Familie jüngster Spross.

Mein Vater grüsst auf's Herzlichste, ebenso mein Bruder Heinrich!
Vom 8ten Januar an bin ich wieder als Docent in Jena thätig.

[5] Carl August Dohrn an von Baer
2.10.1871, Stettin

Stettin 2 October 1871.

Hochverehrter Freund und Gönner

Im Auftrage meines jüngsten Sohns Dr. Anton, bisher Privatdocenten an der Universität Jena, beehre ich mich, Ihnen den beifolgenden Plan seines neuen Unternehmens zu überreichen. Wenn es zu Wasser wird, so ist es wenigstens ein Trost, daß es Salzwasser ist, wobei vielleicht aus Versehen irgend attische Krystalle mitunterlaufen. Da Anton bereits Gelegenheiten suchte und fand, mit zoologischen Celebritäten (Huxley, Darwin, van Beneden, Leuckart, Dubois etc etc) über die Sache zu reden, da sie sammt und sonders sich in hohem Grade befriedigt über Idee und Modalitäten äußerten, so wage ich zu hoffen, daß auch Sie, mein hochverehrter Freund, wohlwollende Theilnahme dem anscheinend kühnen Plane nicht versagen werden. Die Kühnheit besteht weniger in der Idee, aber desto mehr in der realen Schwierigkeit, Napoletaner davon zu überzeugen, daß man fähig ist, an einen rein wissenschaftlichen Zweck Mühe und "baares Geld!!!" ins Ungewisse zu riskiren. Bürgermeister und Schöffen haben sich nicht entblödet, geradezu auszusprechen, daß es mit dem sogenannten Aquarium eigentlich doch wohl bloß auf ein verkapptes hôtel wenn nicht gar bordell gemünzt sei.

Anton ist in der letzten Zeit beständig auf der Landstraße gewesen, hat die neuen (verkehrten) Einrichtungen in Brighton (wo auch ein bassin für einen Walfisch (sic) bereit steht) die Naturforscher Versammlungen in England, Schottland, Rostock besucht, Contracte mit den Lieferanten von Maschinen, Glastafeln, etc vorbereitet, und ist jetzt auf dem Wege zur archäol. Versammlung in Bologna, um dann die Sache in Napoli in Angriff zu nehmen.

Da ich glaube, daß der letzte Krieg für einige Jahre als Sicherheits Ventil ein großes Quantum nephitischer Gase hat entweichen lassen, so hege ich die Hoffnung, daß die finanziellen Berechnungen nicht allzu sanguin aufgestellt sind, und daß wenn einige ruhige Jahre die zoologische Station erst in Gang gebracht haben, sie alsdann für sich selber Zeugniß ablegen kann und wird.

Mein ältester Sohn, Dr Heinrich, welcher die Ehre hatte, sich Ihnen in Petersburg vorzustellen, und der seit 1863 in dem damals fundirten Pommerschen Museum die Zoologica dirigirt und pro virili fördert, wird heute oder morgen von einem Ausfluge nach den United States zurück erwartet, wo ihn Agassiz und Dr H. Hagen in Cambridge freundlichst aufgenommen und so reich für besagtes Museum beschenkt haben, daß er nicht alles zu placiren vermochte.

Auf diese Art documentirt, wie Sie sehen, die Familie D. unvermuthet viel Bestialisches – eheu!

Mir persönlich geht es Gottlob wohl, und ich habe den herzlichen

Wunsch, daß es Ihnen und den Ihrigen ebenfalls so gut gehe. Vielleicht erfreuen Sie durch ein Paar freundliche Zeilen

Ihren Sie herzlichst verehrenden
Dr. C. A. Dohrn.

---

[6] Dohrn/CAD an von Baer
2.1.1873, Stettin

Stettin. 2ten Januar 1873.

Als ich gestern in Ihrer Autobiographie den Abschnitt las, Excellenz, wo Sie von dem Beginn Ihrer embryologischen Studien unter Doellinger erzählen, dacht' ich, es müsse Ihnen doch wie eine Fortsetzung Ihrer eigenen Absichten und Entwürfe erscheinen, wenn gegenwärtig in Neapel bereits die letzte Hand an die Vollendung eines grossen Gebäudes gelegt wird, das eigentlich nur aus dem Bedürfniss entstanden ist, der embryologischen Forschung weitere Mittel und Wege zu ebenen. Sie wissen, Excellenz, wie gering unsre Kenntnisse über die Entwicklung der Fische sind, dass die Embryologie der Cephalopoden, der Würmer, der Echinodermen,–kurz fast aller marinen Thiere erst noch zu machen ist, wennschon grade die letzten Jahrzehnte mehr nach dieser Seite gefördert haben, als frühere Zeiten.

Die grossen Schwierigkeiten, welche diesen wichtigen Untersuchungen aus so vielerlei Ursachen erwuchsen, glaube ich zum grossen Theile durch die Erbauung und Einrichtung der "Zoologischen Station" wenn nicht beseitigt, so doch sehr verringert zu haben, und zugleich habe ich danach gestrebt, eben durch die ganze, der Zoolog. Station zu gebende Organisation auch dafür zu sorgen, dass ihre Benutzung so vielen Zoologen, als irgend möglich zu den geringsten Kosten ermöglicht wird. Ich bin im Begriff, mit den Deutschen Regierungen Abmachungen zu treffen, welche auf der einen Seite der Zoolog. Station verfügbare Mittel, auf der andern den deutschen Zoologen grosse Erleichterungen für ihre Forschungen in Neapel gewähren sollen, und wenn dies für Deutschland geleistet ist, werde ich es auch für die übrigen Länder und vornehmlich auch für Russland zu erreichen suchen, das ja so viele und so tüchtige Kraefte grade nach Neapel sendet.

Ich hätte freilich gewünscht, Ew. Excellenz dies Alles bereits als definitiv beendet zu melden,–aber Excellenz werden mir gestatten einen zweiten Brief an Sie zu richten, worin ich das hoffentlich nachholen kann. Da ich aber zum Schluss des Jahres meine Eltern besuchte und mein Vater mich fragte, ob ich Ihnen schon über den Gang meines Unternehmens Mittheilung gemacht hätte, so verband sich diese Frage mit der eigenen innern Aufforderung, Ihnen wenigstens einen kurzen Bericht zu geben, dass die ganze Sache ihrer Vollendung entgegengeht.

Sie werden, Excellenz, einem jungen Adepten dies nicht für Zudringlichkeit auslegen, sondern den Wunsch erkennen, dem alten Meister zu sagen, wie sehr sich die jetzige Generation als seine Nachkommenschaft in der Wissenschaft empfindet, und wird es auch schwer, ja unmöglich sein, so Grosses und Bahnbrechendes wieder zu leisten, so[b] wird doch "Viribus unitis" auf dem gebrochenen Wege weiter fortgegangen und dem Ziel entgegengestrebt, dass Sie uns gesteckt haben.

Nehmen Sie, Excellenz, diesen Brief als den Ausdruck des Wunsches hin, dass es mir vergönnt sein möge, Ihnen noch recht oft Neujahrs-Berichte über das Ergehen und über die Thätigkeit der Zoologischen Station abstatten zu können; das ist mein und aller Zoologen herzlichster Wunsch.

Mein Vater lässt Sie auf das Herzlichste grüssen, eben so empfiehlt sich mein Bruder Heinrich.

In Ehrerbietung und mit den besten
Wünschen für das neue Jahr
Ihr
ergebner
Anton Dohrn

[CAD:]
Mit den allerherzlichsten Grüssen und Wünschen von Ihrem ergebensten, der im vorigen Jahre aus unbekannten, unermittelt gebliebenen Gründen in Florenz beinah ins Gras gebissen hätte, aber Gottlob bereits wieder solito more vorn und hinten ausschlägt, Schrittschuh läuft und dergleichen juvenilia mit oder ohne Grazie betreibt. Die noble Passion für das lausige Ungeziefer blüht nach wie vor, Sendungen fliegen von und nach Sidney, Neu Guinea, Hottentottenland, Cambridge-Massachusetts [?] kurz die sechsbeinige Freimaurerei steht in vollem Flor.

Dass ich mich für Anton's Unternehmen in jeder Beziehung, sowohl mit dem Beutel wie mit wissenschaftlichen guten Hoffnungen interessire, ist natürlich. Auch Dr Heinrich, der die Ehre hatte, sich Ihres persönlichen Wohlwollens rühmen zu dürfen, setzt seine zoologischen Studien als Director der bestialischen Abtheilung des pommerschen Museums fort, wenngleich er als Stadtrath loci augenblicklich für neue Hafen und DocksAnlagen praeoccupirt ist.

Aus voller Seele wiederholend, dass Ihnen das neue Jahr Heiterkeit und Gesundheit ungetrübt bewahre

Ihr herzlichst ergebner
C. A. Dohrn

---

[b] Corrected with different ink.

[7] von Baer an Dohrn
15./27.1.1873, Dorpat[c]

Dorpat. 15/27. Jan. 73.

Hochgeehrtester Herr Dr. Anton Dohrn!

Ihr geehrter Brief vom 2. Januar kam leider in meiner Abwesenheit an. Da unser Jahr um 12 Tage verspätet sich dem Ihrigen nachschiebt, war ich auf dem Lande, um das Weinachtsfest bei meinen Kindern und Grosskindern zu feiern. Nach meiner Rückkunft fand ich ihn vor; allein, da ich durch eine Verkaeltung auf dieser Winterreise, oder sei eine andre Veranlassung dazu gekommen, ich verfiel alsbald darauf in eine ernstliche Krankheit und bin jetzt nicht nur aus dem Bette, sondern in ziemlicher Restauration.

Ich glaubte das Gesagte vorausschicken zu müssen, um die Verspaetung meiner Antwort verstaendlich zu machen. Zum Inhalte Ihres Briefes übergehend, muss ich vor allen Dingen sagen, dass er auf mich einen überwaeltigenden Eindruck ausgeübt hat. Als ich vor laengerer Zeit von Ihrer grossartigen Unternehmung hoerte, glaubte ich kaum, dass sie zur Ausführung kommen würde. In dieser Stimmung war ich noch, als ich die Ehre hatte, Ihren ersten Brief zu erhalten. Seit jener Zeit aber habe ich so viele oeffentliche und private Nachrichten von der wunderbaren Ausführung Ihres grossen Planes erhalten, dass ich an der Durchführung nicht mehr zweifle und eine neue Aera für die naturhistorische Untersuchung darin zu erkennen glaube. Wenn an einem so reichen Meeresbette, wie das neapolitanische ist, die Mittel der Untersuchung angehäuft, die Methoden derselben ausgebildet und traditionell gemacht werden, so müssen die Erfolge wohl viel reicher und bleibender ausfallen, als wenn, wie es früher geschah, mit wenig Vorbereitung und beschraenkten Apparaten, ohne Kenntniss der Lokalitäten, ja ohne Kenntniss der passenden Zeiten Naturforscher aus weiter Ferne dahin kommen. Ich will darum auch nicht verhehlen, dass ich ursprünglich nicht ohne Neid an Ihre Unternehmung denken konnte, war ich doch vor fast 30 Jahren nach Triest gekommen, um die Entwicklung der Seeigel zu untersuchen, lange vor der Laichzeit dieser Thiere, da ich auf keine Weise die Jahreszeit für dieses Geschaeft auffinden konnte und musste ich doch zurück, als das Laichen begann, weil man in Russland durchaus voraussagen muss, wenn man fertig sein will.

Ich theile ganz Ihre Meinung, dass noch ausserordentlich viel zu thun ist, namentlich für die Entwicklungsgeschichte, und diese ist nicht nur die Leuchte für die Zoologie, sondern das Verstaendniss des Thierlebens selbst. Leider hat mich mein Auge soweit verlassen, dass es mir nicht mehr den Blick in den feineren Bau gewährt, und ich muss mich damit

[c] Copy by Anton Dohrn, original letter not preserved.

begnügen, dass sich mir die Richtung für die Bewegung in meinem Zimmer zeigt. Dennoch sehne ich mich darnach im Herbste dieses Jahres Ihre Anstalt in Neapel zu sehen, wenn mein Befinden noch erträglich sein wird, denn allerdings mehren sich die Mahnungen, dass das Räderwerk oder richtiger wohl das Fadenwerk des Leibes sich abgenutzt hat. Ich bin meinem 81sten Geburtstage nahe und das ist ein Zeitmaass, das schon über das hoehere Normalmaass des Lebens hinausgeht.

Sie Selbst, geehrter Herr Dr., haben so gründliche und tüchtige Arbeiten im Fache der Entwicklungsgeschichte geliefert, dass Sie wohl einer Empfehlung nicht bedürfen, um für Ihre Anstalt die nöthigen Unterstützungen der Regierungen zu erhalten. Diese Anstalt empfiehlt sich meiner Meinung nach, hinlänglich selbst. Doch würde ich mich sehr freuen, wenn ich irgend etwas dazu beitragen koennte, diese Einrichtung ins Leben zu rufen und deren Wirksamkeit einzuleiten, obgleich ich sie vor 50 Jahren noch lieber haette werden sehen. Ich denke zuvörderst an eine oeffentliche Besprechung in der Deutschen Petersburger Zeitung. Doch möchte ich diese am liebsten bis in die zweite Hälfte des Februar verschieben, da ich hoffen kann, um diese Zeit wieder selbst schreiben zu koennen. Bis dahin ist in unserm nordischen Winter das Licht mir nicht genügend zum Lesen und Schreiben. Wenn Sie bis dahin Ihre neue Auseinandersetzung, auf die Sie mich hoffen lassen, mir zuschicken koennen, würde es mir sehr lieb sein.

Den herzlichsten Dank für Ihre freundliche Erinnerung an meine früheren Versuche, die ich jetzt nur noch als Anläufe zu einer spaeteren Erkenntniss betrachte. Es ist immer erfreulich, wenn man diese rudimentalen Versuche nicht mit den spaeteren gründlicheren vergleicht. Auch ich wünsche für Sie und Ihre Anstalt, dass Sie oft noch ein neues Jahr erleben.

Mit ebensoviel Hochachtung für Ihre embryonalen Arbeiten, als Bewunderung Ihrer Ausdauer in der begonnenen Unternehmung

Ihr ergebenster Diener
Dr. K. E. v. Baer.

---

[8] Dohrn an von Baer
8.2.1873, Neapel

Napoli. Palazzo Torlonia. 8.2.1873.

Wenn es überhaupt einer besonderen Anerkennung werth wäre, Excellenz, was ich hier gethan habe, so hätte ich die grösste, die ich überhaupt erhalten könnte, in Ihrem Briefe gefunden. Dass die Zoologische Station durch Ihren Brief noch unmittelbar angeknüpft worden ist an die grosse geistige Entwicklung, in deren Mittelpunkt Sie stehen, dass so dies Institut, das die Kraefte der Gegenwart und der Zukunft fest zusammenfassen und auf die Lösung der gewaltigen Probleme hinarbeiten soll, noch von dem Manne willkommen geheissen wird, der die Traditionen Caspar Friedrich Wolffs aufnahm und so mit seiner eignen Geschichte und Entwicklung in die Entstehungszeit der gesammten modernen Biologie hineinreicht [?]–das hat für mich einen unvergänglichen Werth,–und der Brief, der mir diese Gewissheit brachte ist ein Adelsbrief, der mir das "noblesse oblige" immer zurufen soll, sollte ich je in Gefahr gerathen, meinen Blick von so hohen Vorbildern abzuwenden.

Dass es aber mit meinen Verdiensten nicht so steht, wie Sie es in so freundlicher Weise sagen, das müssen Sie mir schon aufs Wort glauben. Wenn das Schicksal einem jungen Manne gestattet sich nach innerstem Drange auszuleben,–so ist das wahrlich nur ein Verdienst des Schicksals, –und in meinem Falle wird das noch concreter, es wird ein reines und grosses Verdienst meines Vaters, der mir die Mittel gab, durch die allein ich in den Stand gesetzt worden bin zu thun, was ich gethan habe. Dass es manchmal auch Schwierigkeiten gab, die aus der Neuheit der Sache, den eigenthümlichen Zustaenden der Stadt Neapel, und besonders meiner eignen Unerfahrenheit resultirten, brauche ich freilich nicht hinzuzusetzen. Jetzt indess scheint Alles so glücklich überwunden, dass ich mich dem vollsten Zutrauen überlasse, auch in der Zukunft werde es gut und fruchtbar vorwärts gehen,–besonders wenn mir Diejenigen nicht ihre Hilfe versagen, die am besten den möglichen Werth des zu Gewinnenden abzuschätzen verstehen.

Ich erlaube mir, Ihnen eine kleine Broschüre mitzutheilen, welche in den Preussischen Jahrbüchern August 1872 zuerst erschien. Darin steht so mancherlei von dem, was Aufschluss geben kann über Zweck und Ziel, Grund und Ursache,–kurz über die Causalität und Teleologie der Zoologischen Station. Dazu erlaube ich mir folgende weitere Zusätze zu senden, die in einem, der Materie nach von mir selbst herrührenden Aufsatze des Dr. Eisig in der Wiener "Deutschen Zeitung" erschienen sind.

Dann füge ich noch brieflich hinzu, dass nach längeren Verhandlungen mit dem Deutschen Reich, dem preussischen Unterrichtsminister und der Berliner Akademie, Deutschland bereit ist mir 10,000 [Thaler] Subvention zu den Baukosten zu zahlen, dass ferner nach Intervention des Kronprinzen von[d] Preussen *drei* Arbeitstische in den Laboratorien der Station mit jährlich 1500 [Thaler] (à 500 [Thaler]) miethen wird, dass Einleitungen getroffen sind, um 1 Arbeitstisch von Sachsen, 1 von Bayern, 2 von Italien mit denselben Summen (500 [Thaler] = 2000 francs) miethen zu lassen.

Ich würde mich nun an Sie wenden, Excellenz, um zu hören, ob Russland nicht gleichfalls 2 Arbeitstische erwerben will. Solch "Tisch" ist die Collectiv-Bezeichnung für die Darreichung aller derjenigen Utensilien und Vorraethe, welche *ausser* dem Microscop (letzteres exclusive) von

---

[d] "von" has been added by Dohrn in a second moment; "Preussen" referred at first to the State and, after the correction, to the Crown Prince, while the verb has not been adjusted accordingly into "gemietet werden."

einem Zoologen am Meere gebraucht werden, also Chemikalien, Skalpelle, Nadeln, Zeichen- und Schreibmaterialien, *Versuchs-Aquarien, fortdauerndes Material* an lebenden Thieren, Eiern, Larven, grosse, bald sehr vollständige *Bibliothek*, und,—was gewiss sehr wichtig ist,—eine erfahrene wissenschaftliche Umgebung, die ermunternd, helfend, rathend wirkt,—kurzum eine wissenschaftliche Existenz, wie sie *ausserhalb* einer Universität wohl selten zu finden, innerhalb oft schwer genug zu beschaffen, noch schwerer zu bewahren bleibt.

Ich will auch Oestreich-Ungarn 2 Tische anbieten, der Schweiz einen, und falls England, America und andre Staaten oder Privatleute noch solche Erwerbungen machen wollen, so ist noch reichlich Platz vorhanden, die Bedürfnisse zu befriedigen. Auch Frankreich werde ich dasselbe Anerbieten stellen, gewiss aber abschläglich beschieden werden, um so mehr als Herr Lacaze Duthiers sich bereits sehr ungeziemend gegen mich benommen hat,—was er zu seinem Vergnügen auch fürderhin thun kann.— —

Diess gedachte ich Ew. Excellenz mitzutheilen, und da Sie so freundlich waren, Ihre Protection der Zoolog. Station in Aussicht zu stellen, so liegt jetzt der Wunsch vor Ihnen, dessen Erfüllung nicht nur der Station sondern auch andern zu Gute kommen wird. Möchte es Ihnen [?] gelingen, ihm in Petersburg eine günstige Aufnahme zu bereiten.

Dass wir nun noch die Ehre haben sollen, Sie, Herr Geh. Rath hier zu sehen, und das neue Institut von Ihnen persönlich gut[. . .][e] zu lassen, ist fast mehr, als im kühnsten Augenblick gehofft werden kann. Aber wenn es wirklich zur Ausführung kommen sollte, hoffe ich, Ew. Excellenz werden es mich bei Zeiten wissen lassen, damit ich Sie schon an der Grenze meines neuen Vaterland[es] empfangen und Ihnen die Reise durch den italienischen Stiefel nach Kräften erleichtern könne. In dem Wunsche nach solchem Ereigniss vereinigt sich mit mir Ihr specieller Landsmann Dr. Nicolaus Kleinenberg aus Libau, der Verfasser der kürzlich erschienenen Monographie über Hydra, mein treuer Freund, der es vorgezogen hat, vereint mit mir die Zoologische Station zu entwickeln und der darum die Deutschen Universitäten, wie ich, verlassen hat. Er bittet mich, Ihnen ganz besonders seine verehrende Anhaenglichkeit auszudrücken.

Und nun verabschiede ich mich für heute in der Hoffnung, dieser Brief möge Sie in frischer Gesundheit finden, und uns und dem Archive der Zoolog. Station noch weitere Mittheilungen aus des alten, verehrten Meister's eigner Hand bringen.

Gönnen Sie der Jugend, Excellenz, dem Alter dankbar die Hand zu drücken.

In treuster Gesinnung
Ihr
Anton Dohrn

---

[e] One word lost through pasting; "gutheissen" seems to be too long.

[9] von Baer an Dohrn
18.8.1873, Dorpat

Geehrter Herr Doktor!

Ich bin in St. Petersburg gewesen. Ich fand daselbst Ihr Schreiben an das Ministerium vor; es war an die Akademie der Wissenschaften zur Begutachtung übersendet. Die Akademie hat aber bis zum 15ten August a.St. (27.Aug. N.St.) Ferien und ich fand daher auch fast keinen Akademiker gegenwärtig. Mit dem Sekretaire der Akademie besprach ich umständlich die Sache. Auch entwarf ich auf den Wunsch des Sekretaires ein Gutachten von meiner Seite, in welchem ich besonders hervorhob, daß für kein Land Ihre Anstalt größere Wichtigkeit habe, als für Rußland, da von allen Meeren, welche das Europäische Rußland bespülen, nur das Eismeer eine Mannigfaltigkeit von Thieren ernährt, am Eismeere aber keine Stadt, ja nicht einmal ein gut erleuchtetes Haus sich findet, in welchem man mikroskopische Untersuchungen anstellen könnte. Die Ostsee ist bekanntlich östlich von Rügen mehr ein Landsee mit etwas brakischem Wasser, das Schwarze Meer ist nicht viel mehr rei[c]her und das Kaspische Meer noch ärmer. Wir schlossen damit, das [*sic*] das Russische Reich wenigstens auf 2 Tische pränumeriren müßte und hatten die Zeitbestimmung vorläufig auf 10 Jahre festgesetzt. Allein da bei Durchsicht Ihres Antrages es sich fand, daß Preußen und andere Staaten nur auf 3 Jahre subscribirt hatten, so meinte der Secretaire, daß man fürs Erste über dieses Zeitmaß nicht hinausgehen könnte. Ich zweifle gar nicht, daß die versammelten Akademiker wenigstens auf diese Proposition dringen, oder mehr proponiren werden. Herr Owsjannikow war noch nicht zurück, wird aber wohl Ende August eintreffen, und seine Stimme wird ohne Zweifel sehr von Gewicht sein, da die Russische Regierung mit Unterstützung der Wissenschaften nicht geizt, aber auf die Wünsche von Nationalrussen doch noch mehr Gewicht legt, als auf die der Deutschen.

Zum Schluße unseres August oder in den ersten Tagen des September gedenke ich wieder nach St. Petersburg zu reisen, und da werde ich wohl auch an diesen Verhandlungen theil nehmen; denn wenn ich da bin, habe ich wieder alle Rechte eines Akademikers. Ich theile Ihnen dieses mit für den Fall, daß Sie noch einiges über Ihre Anstalt zu sagen hätten. Ich bitte diesen Nachtrag dann mir zuzuschicken. Wenn ich nicht im Stande sein sollte zu reisen, was bei meiner jetzt sehr schwankenden Gesundheit wohl möglich ist, so würde ich diese nachträgliche Nachricht brieflich hinschicken. Vor dem 12ten September nach Ihrem Styl, werde ich schwerlich abreisen, und wie ich hoffe, auch nicht nach dem 16ten.

Den Bericht des Herrn Eisig in der Wiener Deutschen Zeitung 72 No. 344, welchen Sie so gütig waren, mir mitzutheilen, werde ich nach St. Petersburg mitnehmen. Vielleicht können Sie mich noch auf andere, neuere Berichte verweisen.

Mit dem herzlichen Wunsche, daß alle Ihre für die Wissenschaft so erspriеslichen Unternehmungen gedeihen mögen, habe ich die Ehre mich zu zeichnen als Ihren

ganz ergebenen
Dr K E v Baer

Dorpat d. 18. Aug.
1873.

---

[10] Dohrn an von Baer
7.9.1873, Brighton

Excellenz!

Anbei erlaube ich mir, Ihnen zwei Photographien des Zoologischen Palais' in Neapel zu überreichen, und hoffe, Sie möchten daraus von Neuem die Lust schöpfen, uns dort unten im Winter, wenigstens im nächsten Frühjahr einen Besuch abzustatten.

Es hat noch grosser unablässiger Anstrengung bedurft, um die täglich auftauchenden Schwierigkeiten zu bewältigen; und an unvorhergesehenen Unglücksfällen,—wie z.B. dem Platzen von 10 grossen dicken Glasscheiben im Werthe von 2000 Thalern,—hat es leider auch nicht gefehlt.

Das Schlimmste war, dass seit dem Januar meine Kraefte anfingen zu ermatten, und dass die nervöse Abspannung einen so hohen Grad erreichte, dass ich gänzlich von aller Arbeit mich zurückziehen und im Juni Neapel verlassen musste, da die Hitze unerträglich ward. So habe ich 6 Wochen in der Schweiz, 14 Tage in Baden bei meiner Schwester und seitdem 3 Wochen in England an der Seeküste zugebracht, um mich zu erholen.

Dadurch sind natürlich viele neue Weitläuftigkeiten entstanden und so wird sich denn auch die definitive Fertigstellung der Station wohl bis zum Ende des Jahres verzögern.

Darf ich Sie, Herr Geheime Rath, nun wohl mit einer Bitte belästigen? Ich habe vor Monaten die Russische Regierung gefragt, ob sie 2 Arbeitstische in den Laboratorien der Station miethen wolle, nach dem Vorgange von Preussen, Italien, Baiern, Baden, Elsass, Holland und der Universität Cambridge. Die Moskauer Gesellschaft der Naturforscher hat diesen Vorschlag dringend zur Annahme empfohlen; darf ich wohl hoffen, dass Sie, Excellenz, in einem directen Briefe an den Grafen Tolstoi mein Gesuch empföhlen? Ich weiss, die Dinge gehen langsam in Petersburg,—aber gerade darum wird es gewiss sehr vortheilhaft sein, wenn auch von Ihrer Seite dem Minister zugeredet wird.

Und die Station bedarf dringend solcher Unterstützung. Agassiz hat in einem Jahre 300,000 Thaler geschenkt erhalten,—aber in Deutschland kennt man americanische Freigebigkeit nicht. Und ich habe immer mit Geldnoth zu kämpfen.

Und noch eine zweite Bitte. Gewiss gibt es eine Marmor- oder Gypsbüste von Carl Ernst v. Baer; darf die Station wohl hoffen, dass sie eine Copie derselben in dem Laboratorium aufstellen kann?

Mit diesen beiden Fragen und Bitten endet heute meine Zudringlichkeit, für die ich aber um so mehr auf Absolution rechne, als ich sie nur in meiner Eigenschaft als Zoologischer-Stations-Gründer beging und dies Amt eine gewisse Zudringlichkeit als Conditio sine quâ non des Gelingens erfordert.

In der Hoffnung, dieser Brief möchte Sie in vollem Wohlsein treffen und mir durch die Stettiner Adresse meines Vaters ein Paar Zeilen als Antwort eintragen, grüsse und empfehle ich mich

in schuldiger Anhaenglichkeit
und Ehrerbietung
Ihr
ergebner
Anton Dohrn

Brighton 7. September 1873.

---

[11] von Baer an Dohrn
31.8 /12.9.1873, Dorpat

An
Herrn Doctor Anton Dohrn.

Hochgeehrter Herr Doctor!

Ihren Brief aus Brighton vom 7. Sept. 1873 habe ich gestern Abend erhalten & ich eile, ihn sogleich zu beantworten.

Da muss ich vor allen Dingen sagen, dass ich Ihnen vor etwa 12 Tagen unmittelbar nach Neapel geschrieben habe & ich wundre mich, dass man Ihnen diesen Brief nicht nachgeschickt hat. Vielleicht ist er aber auch gar nicht angekommen, da ich Ihre Adresse nicht genau kannte. Ich quälte mich, eine genügende italiänische Adresse zu finden und schrieb Napoli nel nuovo Instituto [*sic*][f] Zoologico. In diesem Briefe theilte ich mit, dass ich in St. Petersburg gewesen war und dort die Sache wegen Praenumeration auf zwei Tische für russische Beobachtung im Gange fand. Der Unterrichtsminister, Graf Tolstoi hatte diese Angelegenheit der Academie zum Begutachten übergeben; die Academie hatte aber Ferien bis zur Mitte des August, erst am 28. August nach unserem Stil konnte die Sache vorgenommen werden. Ich erklärte natürlich mein warmes Interesse dafür. Der Secretär meinte, es würde gut sein, wenn ich ein oeffentliches Zeugniss abgäbe. Das habe ich denn auch gethan, obgleich

---

[f] Correct: Istituto.

nach meiner Meinung, diejenigen Academiker, die kürzlich in Neapel gewesen sind, es besser hätten thun können, und die Stimme eines Nationalrussen bei dem Minister mehr gewirkt haben würde. Allein, da ich jedenfalls ganz unparteiisch bin, weil ich die Vortheile nicht mehr geniessen kann, da nicht nur 82 Jahr auf mir lasten, sondern auch meine Augen mir den Dienst versagen, so wird man meine Stimme wenigstens, wie die eines Unparteiischen ansehen müssen.

Ich habe daher in die Petersb. Zeitung einen Artikel eingerückt, auf den die Academie sich berufen wird, und den ich hier beilege. Sie haben Recht, dass die Angelegenheiten in Russland langsam gehen, in diesem Falle aber war es vorauszusagen. Es hat nämlich der Minister des Unterrichts über gar kein Geld zu verfügen, das nicht schon seine Bestimmung hätte. Er muss daher einen Antrag an den Finanzminister stellen und um eine besondere Bewilligung bitten. Um das thun zu können, braucht er einen nachdrücklichen Antrag von einer dazu berufenen Behörde, daher die Anfrage bei der Academie. Ich zweifle gar nicht, dass diese nachdrücklich ausfallen wird, doch habe ich den Secretär gebeten, wenn sich Schwierigkeiten finden sollten, mir es mitzutheilen. Dann schreibe ich eine zweite Erklärung, deren Inhalt sein soll: "какъ вамъ не стыдно" d.h. Schämt ihr euch nicht! und ich denke diese Expectoration von einem bald Verschwindenden soll nicht ohne Wirkung bleiben. Ich zweifle nicht, dass vor Beendigung des Jahres man zwei Tische für Beobachter aus Russland wird bestellt haben, doch wird die Bestellung wahrscheinlich erst mit dem nächsten Jahre beginnen. Ich hatte mit dem Secretär schon abgemacht, dass man auf 10 Jahr subscribiren sollte, allein, da derselbe fand, dass Sie selbst in dem Schreiben an den Minister berichtet hatten, Preussen und andre Staaten hätten auf drei Jahr subscribirt, so hatte er auch nicht Lust, weiter zu gehen.

Sie dürfen mir vertrauen, dass es wirksamer ist, wenn ich öffentlich spreche, als wenn ich an den Minister Tolstoi schreibe. Ein solches Schreiben kann er in den Papierkorb werfen, eine öffentliche Erklärung nicht. Soviel glaube ich zu gelten, dass man ihm eine solche auch künftig vorhalten werde.

Herzlichsten Dank für die Photographie, die ein Prachtgebäude darstellt.—Eine Marmor- oder Gipsbüste von K[g] E. v. Baer existirt aber nicht, und würde jetzt auch wenig erbaulich sein, da sie nur einen Abgängigen darstellen könnte.

Schliesslich bitte ich um eine sichere Adresse nach Neapel.

Mit vollkommenster Hochachtung
Ihr ergebenster
Dr K E v Baer

Dorpat, d. 31. Aug. 1873.
(12. Sept.)

---

[g] Originally "C," corrected into "K."

[12] Carl August Dohrn an von Baer
18.9.1873, Stettin

Stettin 18 Septb 1873

Lieber verehrter Freund und Gönner

Herzlichen Dank für Ihr willkommenes vom 12 d.—ich komme so eben von einer Reise nach Carlsruhe zu meiner Tochter zurück, und konnte deshalb nicht eher antworten.

Dr. Anton soll Ihren, ihm gewiss hocherfreulichen Brief sammt dem Inserat erhalten (—Renard schickt mir ebenfalls die Petersb. Zeitung—) aber ich werde ihm beides wohl nach England schicken, wenn er wie ich vermuthe, noch dort ist, um sich bei Hrn Lloyd, Dirigenten des Aquarium im Crystal Palace, noch einige feinere technica bei der Manipulation des Röhren Apparates einpauken zu lassen.

Es sollte mich und natürlich auch Anton sehr freuen, wenn die Royal Society durch Ihre vis prophetica sich gedrungen fühlte, Ihre Verheissung zu erfüllen. Uebrigens haben viele und achtbare Gesellschaften schon Brillantes eingesandt.

Dass *Ihr* Eintreten für Anton nicht blos für russische sondern auch noch für *andre* Finanz Minister von erheblichem Einfluss ist, davon bin ich gründlich überzeugt.

Leider ist einer von Anton's Förderern, Prof. Czermak, eben seinem mehrjährigen Diabetes erlegen. Er war erst 45 Jahre alt! Schade um den tüchtigen Mann. Aber da auch Patroklus sterben musste—

Einstweilen wollen wir aber Diis faventibus die Sache noch weiter mit ansehen!

In alter Anhänglichkeit
Ihr
C. A. Dohrn.

Den Argonauten Heinrich erwarten wir in den nächsten Tagen von New York.

[13] Dohrn an von Baer
19.9.1873, London

London. Charing Cross Hotel 19. Sept. 73.

Excellenz!

Ich eile Ihnen noch zu sagen,—dass fast nur Preussen auf 3 Jahre gemiethet hat, die andern Regierungen auf 5, und dass es mir viel lieber ist, wenn Russland über Preussen hinausgeht,—denn dann kann ich Preussen treiben, und werde es auch jetzt thun.

Holland hat einen Tisch gemiethet, auch Oxford trifft Anstalten.

Mit grösstem Dank für Ihre Theilnahme und in schuldiger Verehrung

Ihr
ergebner
Anton Dohrn

Darf ich Sie ersuchen, für die Zoolog. Station die sämmtlichen biologischen Schriften der Kaiserlichen Akademie der Wissenschaften zu erbitten. Bereits haben die Akademieen von Kopenhagen (67 Bände) Berlin u. Neapel ein Gleiches gethan. Royal Society ist im Begriff, und ein [*sic*] grosse Zahl Andrer weniger grosser Akademien und Gesellschaften folgen dann.

In grosser Eile

Ihr AD.

Ich werde Mitte October wieder in Neapel sein.

---

[14] Dohrn an von Baer
29.9.1873, Stettin

Stettin. 29. Sept. 1873.

Excellenz!

Gestern bin ich hier in meiner Heimath angekommen und beeile mich Ihnen für den zweiten Brief und die Uebersendung des Artikels in der St. Petersburger Zeitung meinen herzlichsten Dank zu sagen.

Ihr erster Brief nach Neapel hat mich kurz vor meiner Abreise in London getroffen, zusammen mit einem andern, der mir von Seiten eines frühern Mitgliedes des Unterrichts-Ministerium in St. Petersburg zugesandt wurde und dieselbe frohe Botschaft enthielt, wie Ihr Brief.

Was nun die Station anlangt, so kann ich nur sagen, dass sie vorzüglich gedeiht, dass alle die unzähligen Schwierigkeiten langsam aber sicher aus dem Wege geräumt werden und dass die ganze Angelegenheit so populär geworden ist wie lange kein wissenschaftliches Unternehmen.

Bei meinem diesmaligen Aufenthalt in England ist es mir wieder gelungen eine Reihe von Vortheilen dem neuen Institute zu erwerben. Die englische Naturforscher-Versammlung hat mir 200 Thaler zugewiesen behufs Anstellung von Experimenten zur Zucht von Eiern und Larven niederer Seethiere, Oxford wird, so hoffe ich, der Schwester-Universität Cambridge folgen und einen Tisch miethen, die Royal Society hat einen grossen Theil ihrer Schriften geschenkt (so hat also Ihr Aufsatz in der St. Petersburger Zeitung die Wahrheit anticipirt) und eine grosse

Menge von andern Gesellschaften folgt diesem Beispiel, – die Bibliothek wächst also *sehr* beträchtlich und rasch.

Da nun auch die Akademie von Berlin, die von Kopenhagen, Neapel ecc. ecc. ihre Schriften geschenkt haben, so werde ich mich demnächst auch an die St. Petersburger Akademie mit gleichem Gesuch wenden, und hoffe dann auf keine abschlägliche Antwort zu stossen. –

Es wird Sie gewiss freuen zu hören, dass *Huxley* wieder ganz hergestellt ist. Er war sehr krank, leidet offenbar an organischen Verdauungsstörungen, die sein Nervensystem arg zerrütteten. Ich besuchte ihn vor meiner Abreise und fand ihn im Kreise seiner Familie so lustig und vergnügt wie je. Er trug mir auf, Ihnen die herzlichsten Grüsse zu bestellen. –

Mein Vater wünscht Ihnen die ungetrübte Fortdauer Ihrer wunderbaren Gesundheit, – dass ich und mit mir die ganze Gesellschaft Derer, die sich an die Zoologische Station anschliessen, diesem Wunsche gesellen [*sic*], brauche ich kaum hinzuzufügen.

Ich hoffe, Ihnen im nächsten Jahre berichten zu können, dass die Arbeiten in den Laboratorien begonnen haben, – eine grosse Zahl Deutscher, Englischer, Italienischer, Holländischer Zoologen haben sich bereits angemeldet. –

In herzlichster Ergebenheit und mit aufrichtigstem Danke

Ihr
Anton Dohrn

Meine Addresse in Neapel ist:
Dr. A. D. / Stazione Zoologica.

---

[15] von Baer an Dohrn
18.10.1873, Dorpat

Dorpat, 18ten Oct. '73

Hochgeehrter Herr Doktor!

Ich will nicht unterlassen, Ihnen anzuzeigen, daß die Akademie zu St.Petersburg auf die Pränumeration von 2 Tischen angetragen hat. Daß der Kaiser sie bewilligt, wenn nur der Minister den Antrag weiter befördert, ist nicht zu bezweifeln; doch bleibe ich bei meiner Vermuthung, daß man erst mit dem künftigen Jahre beginnen wird. Ich habe auch von meiner Seite bei der Akademie darauf gedrungen, daß sie Ihnen ihre Biologica zusendet, obgleich ich kaum glaube, daß das nöthig war, besonders wenn Sie ihr geschrieben haben, daß so viele andere Akademien vorangegangen waren. Uebrigens habe ich seit längerer Zeit keine neueren Nachrichten aus St. Petersburg.

Mit dem herzlichsten Wunsche, daß Ihre Gesundheit wieder ganz befestigt werde, was ja zum Gedeihen der neuen Anstalt so nothwendig ist, zeichne ich mich mit ebensoviel Herzlichkeit als Hochachtung

als Ihr ergebenster Diener
Dr K E v Baer

---

[16] Dohrn an von Baer
23.1.1874, Neapel

Stazione Zoologica di Napoli
23. Januar 1874.

Excellenz!

"Doch Alles stumm bleibt, wie zuvor!"

Herr Graf Tolstoi hat nicht eine Silbe verlauten lassen, ob er geneigt oder abgeneigt sei, mein Anerbieten anzunehmen und die Empfehlung der Akademie für mehr zu halten, als für schätzbares Material. Die Akademie mag das nun mit Herrn Tolstoi ausmachen, wie sie will,—das ist ihre Sache, aber meinerseits halte ich es für geboten, dem Herrn Grafen ein Entweder-Oder zu stellen, da von andern Seiten Ansprüche an die Zoolog. Station gestellt werden, die ich ablehnen muss, so lange ich die beiden Tische zur Verfügung der Russischen Regierung zu halten habe.

Ehe ich aber einen darauf hinzielenden Brief an den Herrn Minister schicke,—und der Herr Graf versieht sich wohl des Factums nicht, dass ich ein unabhängiger Mann bin und wohl der Höflichkeit einer Antwort werth,—wollte ich Ihnen erst diese Mittheilung machen, da Sie Sich meiner Unternehmung mit so viel Theilnahme angenommen haben. Vielleicht geben Sie mir einen Rath, was ich am besten thue, um mir das Spiel nicht zu verderben.—

Trotz unermüdlicher Thätigkeit bin ich doch noch nicht mit den gesammten Einrichtungen fertig. Zum Theil liegt die Schuld an einer langwierigen nervösen Erschlaffung infolge zu andauernder Anspannung und gleichzeitiger gemüthlicher Emotionen, dann aber auch an der Complicirtheit der Aufgabe.

Dennoch wird in 3 Tagen das Aquarium, das 60 Bassins enthält, geöffnet, und ich hoffe, das Publicum werde damit zufrieden sein.

Für nächsten Monat sind ein Schwede, drei Englaender und im März zwei Hollaender angemeldet,—also fehlt es nicht an arbeitslustigen Leuten, denen die Station von Nutzen sein wird.

Und zum Schluss dieser kurzen Mittheilungen muss ich Ihnen noch gestehen, dass wir trotz alledem eine Büste von Ihnen in dem Prunksaale der Zoolog. Station aufgestellt haben. Dieselbe ist von dem kürzlich sehr berühmt gewordenen jungen Bildhauer Adolf Hildebrand nach dem Bilde, das in Ihrer Selbst-Biographie sich findet, gemacht,—und wie mir von Jemand mitgetheilt wird, der Sie öfters in Dorpat gesehen hat, sehr aehnlich. Werden Sie dies Attentat auch ratifizieren?

In der Hoffnung, das [*sic*] Sie trotz des Winters doch die alte Rüstigkeit haben, grüsse ich von ganzem Herzen und bin beauftragt von meinen Freunden u. Mitarbeitern an der Station, Dr. Kleinenberg u. Dr. Eisig gleichfalls die ehrfurchtsvollsten Grüsse zu bestellen.

Mein Vater, der in Genova ist, wird sich im nächsten Monat das Gebäude und Alles was drin ist betrachten und hoffentlich finden, dass sein Sohn sein Geld nützlich angewendet hat. Wüsste er, dass ich an Sie schriebe, so würde ein Gruss in seinem Style gewiss nicht fehlen.

Ihr
ehrfurchtsvoll ergebner
Anton Dohrn

---

[17] von Baer an Dohrn
24.1./5.2.1874, Dorpat

Hochgeehrter Herr Doktor!

Ihr geehrtes Schreiben vom 23ten Jan. habe ich vor einigen Stunden erhalten und ich beeile mich eine darauf bezügliche Antwort zu diktiren.–

Der Graf Tolstoi, der sehr angelegentlich bemüht ist, das Studium der alten Sprachen in Rußland zu heben, obgleich die Russen bisher für dieses Studium sehr wenig Sinn gezeigt haben, mag nun glauben, daß es nicht nöthig ist, ihr Studium der Natur zu fördern, weil sie dafür Sinn und Talent haben, oder er mag die Sache blos vertrödelt haben; jedenfalls hat er sehr ungebührlich gehandelt, so lange zu schweigen. Es scheint mir durchaus nothwendig, daß Sie ihn um eine kategorische Antwort bitten, da Sie doch unmöglich Andere abweisen können, um auf ihn zu warten. Ich denke, Sie könnten ungefähr so schreiben:

> Sie hätten dem russischen Reiche vorgeschlagen, auf 2 Tische zu subscribiren, in der Meinung, darin diesem Reiche ein [*sic*] Gefallen zu erweisen und Ihre Achtung zu bezeugen. Da in den letzten Jahren einige junge Naturforscher Russischer Nation sehr geachtete und ausgezeichnete Arbeiten über die Entwickelungsgeschichte von Seethieren geliefert haben und häufig Neapel die Stätte ihrer Untersuchung war, so hätten Sie geglaubt, man müßte die Bequemlichkeiten und Mittel, welche Ihre Anstalt gewährt, auch den Russen anbieten. Sie hätten aber noch keine Antwort erhalten. Nun mache man aber auf die letzten noch unbesetzten Tische von anderer Seite Ansprüche und Sie [hätten] keine Möglichkeit dieselben abzuweisen, solange Sie keine Zusicherung aus Rußland erhalten hätten. Ich würde geradezu einen Termin bestim-

men, bis zu welchem Sie warten wollten, oder nach welchem Sie es aufgeben würden, Russische Naturforscher in Ihrem Institute zu sehen.

Ich glaube, Sie würden für die Zukunft gewinnen, wenn Sie gar nicht bittend auftreten, da man eine kategorische Erklärung verlangt.

Es ist wirklich merkwürdig, daß die jungen Russen gerade in diesem Fache der feineren Beobachtung sich ausgezeichnet haben, obgleich ihre Seeküsten dazu gar nicht einladen und daß die Regierung jetzt gar nichts für diese Studien scheint thun zu wollen und ihnen dagegen eine andere Branche aufdrängt, für welche die Russen nun einmal keinen Sinn haben.

Ich werde auch dem Secretaire der Akademie schreiben, damit dieser weiß, welche Wirkung ihre Empfehlung gehabt hatt [*sic*]. Vor wenigen Tagen noch sagte mir ein hiesiger Naturforscher, daß er nach Neapel zu gehen wünsche, um die Entwickelungsgeschichte des Amphioxus zu revidiren.

Also haben Sie doch meine Büste aufgestellt; wunderbar genug, aber zuviel Ehre für mich. Wenn ich in diesem Sommer noch gehörig zusammenhalte, wünsche ich in der That Ihre Anstalt noch zu sehen, um dann von der Welt Abschied zu nehmen. Allein die Mahnungen von Freund Hain werden immer ernstlicher und im Anfange dieses Jahres bin ich ernstlich krank und sehr schwach gewesen. Es kommt nur darauf an, wie der Uebergang zu unserem Frühling sich machen wird. Jetzt haben wir endlich Winter, der lange auf sich hat warten lassen.

Kürzlich habe ich von Agassiz ausführliche Nachrichten über die Mittel, die man ihm zur Disposition für sein Museum gestellt hat, erhalten. Er hat sie noch auf seinem Todesbette zusammentragen lassen. Ich bin ganz erstaunt über die Großartigkeit dieser Mittel. Welch ein Unterschied zwischen Amerika und Europa!

Ich habe im Sommer des verflossenen Jahres eine Abhandlung drucken lassen, in der ich die Deutung bezweifle, die man der Entwickelung der Ascidien gegeben hat. Da diese Abhandlung in den Memoiren der Akademie zu Petersburg erschienen ist, schien es mir überflüssig, sie besonders zu überschicken; allein, da auch die Akademie häufig dem Prinzip huldigt, "was lange währt, wird gut," will ich doch versuchen, einen Abdruck besonders abzuschicken.

Indem ich Ihnen herzlich Glück wünsche zum Fortschritt Ihrer Unternehmung, habe ich die Ehre mich zu zeichnen als Ihren ergebensten

Dr K. E. v Baer

Dorpat. 24 Jan./ 5 Febr. 74

P. S. Herrn Kleinenberg, dessen vortreffliche Schrift über Hydra ich sehr gut kenne, bitte ich, meine herzliche Empfehlung zu bringen.

B.

[18] von Baer an Dohrn
23.3./5.4.1874, Dorpat

Dorpat. 23 März/5 April. 1874.

Hochgeehrter Herr Doktor!

Zu meinem großen Bedauern muß ich mittheilen, daß wohl nur die Besetzung eines Tisches Ihrer Anstalt von Seiten des Russischen Reiches zu erwarten steht. Der beiliegende Brief des beständigen Sekretairs der Akademie der Wissenschaften zu St. Petersburg, an mich gerichtet, wird Ihnen die nähere Auskunft darüber geben. Ich hatte den Sekretaire gebeten, doch dafür Sorge zu tragen, daß zwei Tische belegt würden; der beiliegende Brief ist eine Antwort darauf. Es ist fast unglaublich, daß der Minister so lau ist, eine Richtung des Studiums zu unterstützen, in welchem Nationalrussen sich mehrfach ausgezeichnet haben, und dagegen die sogenannten "klassischen Studien" einheimisch machen will, in welchen noch niemals ein Russe sich ausgezeichnet hat. Diese ganze Richtung ist, wie es scheint, dem Russischen Geiste zuwider.–Ich werde wohl noch einen kleinen Aufsatz veröffentlichen,–aber er wird wohl vergeblich sein.

Haben Sie meine Antwort auf Ihren letzten Brief und den Aufsatz über "die Ascidien" erhalten. Ich frage nicht ohne Veranlassung. Es hat sich nämlich leider herausgestellt, daß in der letzten Zeit mehrfache Sendungen von Briefen und Banderollen [*sic*], die aus Dorpat abgehen sollten, nicht angekommen sind. Man hat einen "Markensammler" in Verdacht, der hier oder in Riga sein müßte.–

Mit der Hoffnung, daß für das neue Jahr doch die Pränumeration für 2 Tische zu erlangen sein wird, verharre ich noch. Doch bitte ich, mir gelegentlich mitzutheilen, welche Art von Vergünstigungen, Büchersendungen u. dgl. Sie von anderen Seiten erhalten haben. Man liebt es in Rußland, gegebenen Beispielen zu folgen.

Mit der vorzüglichsten Hochachtung
Ihr ergebenster
Dr K E. vBaer

[19] Dohrn an von Baer
16.4.1874, Neapel

Napoli. Stazione Zoologica. 16.4.74.

Hochgeehrter Herr GeheimRath!

Seit gestern bin ich wieder nach Neapel zurückgekehrt, da mich der Rath meines Arztes nach Sorrento verwiesen hatte. Meine Gesundheit hat sich nämlich während des Winters nicht gehoben, sondern eher verschlechtert, und in den letzten 6 Wochen habe ich noch zu meinem andauernden Nervenleiden eine Bronchitis und Malaria-Anfälle gehabt.

Das Alles hat mich so gut wie gänzlich von meiner Thätigkeit und auch von der Correspondenz abgeschnitten, und so habe ich nicht nur von ganzem Herzen Ihnen zu danken für Ihre unermüdliche Hilfe und Theilnahme, sondern auch Ihre Verzeihung zu erbitten für die Verzögerung dieses Dankes und meiner Antwort.

Von der Russischen Regierung ward mir die Anzeige, dass *ein* Tisch gemiethet werden solle. Auch diesen Brief habe ich noch zu beantworten. Derweil habe ich noch von andern Seiten Aufschlüsse über die ganze Sache gehabt, und glaube annehmen zu dürfen, dass Eifersüchteleien zwischen Moskauer u. St. Petersburger Elementen im Spiel gewesen sind. Aber man versichert mich von allen Seiten, dass auf erneute Aufforderung hin, der zweite Tisch auch genommen werden dürfte. Und das ist freilich durchaus wünschenswerth.

Erlauben Sie mir nun, Ihnen folgende Mittheilungen über den Stand der ganzen Sache zu machen.

Die Kosten des ganzen Unternehmens haben sich so vergrössert, dass meine Mittel weit dahinter zurückbleiben, ich also gezwungen ward, von Freunden Geld zu borgen. Das war schlimm genug. Aber zugleich blieb der Ertrag des Aquariums auch hinter gerechten Erwartungen zurück. So stand ich diesen ganzen Winter einer aeusserst schwierigen Lage gegenüber, deren Wirkung auf mein Nervenleben eben auch sehr verhängnissvoll gewesen ist. Schon machte ich mich darauf gefasst, eine Katastrophe zu erleben, als mir Hilfe von England kam. Huxley, mit dem ich seit vielen Jahren in sehr intimen Beziehungen stehe, regte den Gedanken einer Subscription unter englischen Naturforschern und Liebhabern solcher Studien an, und der erste, der zeichnete, war Darwin. Mit einem aeusserst freundlichen und zartfühlenden Briefe sandte er 120 Pfund Sterling, als seine und seiner beiden Söhne Subscription. Man hofft auf diese Weise 1000 Pfund Sterl. zusammenzubringen.

Ein aehnliches Vorgehen hoffe ich auch in Deutschland veranlassen zu können, und werde mich sogar persönlich an den Kronprinzen in Berlin wenden, um ihn zu bitten, Selbst an die Spitze einer solchen Liste sich zu schreiben. Auf andre Weise fürchte ich von reichen Deutschen wenig zu erhalten. Zugleich will ich dann auch noch einen Versuch machen, eine weitere Unterstützung des deutschen Reiches zu erbitten.

Auf diese Weise denke ich das junge Institut schuldenfrei zu machen, und es zugleich immer mehr in das oeffentliche Interesse einzubürgern.

Am liebsten freilich wäre es mir, falls es gelänge, die jährlichen Beitraege der Regierungen so zu erhöhen, dass im Nothfall der Gesammtbetrieb daraus gedeckt werden könnte, für den Fall, dass während eines Krieges oder bei Cholera der Fremdenbesuch in Neapel ausbliebe. Aber ob das gelingen wird, ist mir noch zweifelhaft.

Gegenwärtig arbeiten in der Station 3 Englaender, 2 Russen, 2 Deutsche, 2 Italiener und 1 Hollaender. Ausserdem sind in Neapel noch 5 oder 6 russische Zoologen und in Messina ein Schweizer,—Beweis genug, wie sehr sich die Frequenz gesteigert hat, und wie gut es ist, in der Station allmälig immer mehr einen Centralpunkt zu schaffen, von wo aus für Alle Hilfe und Förderung ausgehen kann.

Aber die Regierungen sollten sich eben sagen, dass dies Unternehmen nur dann diesen Zweck wirklich erreichen kann, wenn es soviel Mittel besitzt, um sicher leben zu können,—und darum ist es eben wichtig, dass feste Jahresbeitraege kommen. Preussen zahlt 1000 [Thaler] hat aber keinen Zoologen geschickt,—aber Minister Falk hat mich versichert, dass dieser Jahresbeitrag nicht wieder gekündigt werden würde,—ebenso Baiern's 500 [Thaler].—

Allerdings habe ich mit herzlichstem Danke Ihren Aufsatz über die Ascidien erhalten und sofort gelesen. Aber ich vermochte damals nicht sofort zu antworten.

Diese grosse Frage beschäftigt ja gegenwärtig alle Zoologen und wenn es auch schwer sein mag, vorauszusehen, wohin nun die Lösung verweist, so scheint doch sicher, dass sie nicht mehr lange auf sich warten lassen wird.

Meine eigne Stellung zu dieser Frage ist eine durchaus isolirte,—sie knüpft an den alten Geoffroy'schen Satz von den auf dem Rücken laufenden Wirbelthieren an,—den Insekten. Durch ein Columbus-Ei glaube ich, hier eine befriedigende Aufklärung geben zu können,—aber es kostet noch so eminent viel Beobachtung, dass ich,—jetzt durch die Station ganz von eigner Arbeit abgedrängt,—vorläufig wenigstens nicht daran denken darf, mich in diesen Streit zu mischen.

Im Sommer will ich meine Insecten-Entwicklungsgesch[ich]ten zum Abschlusse bringen, und hoffentlich erlaubt mir dann der Stand der Zool. Station während des Winters in Neapel ebenso die seit mehreren Jahren begonnenen Untersuchungen der Fisch-Embryologie weiter zu fördern. Derweil werden von allen Seiten so viel Forschungen über Ascidien, Salpen, Myxine, Petromyzon etc. kommen, dass wohl ein grosses Material zu Schlüssen vorhanden sein mag, das Einem erlaubt, etwaige Deductionen zu [veri]fiziren. Mag nur von allen Seiten sine irâ et studio vorgegangen werden!—

Werden Sie es für sehr unbescheiden finden, Excellenz, wenn ich diesem Briefe eine Photographie beifüge, und ganz schüchtern frage, ob es Hoffnung giebt, ein Bild von Ihnen für mich zu bestimmen? Gern würde ich versucht haben, mich persönlich Ihnen vorzustellen, aber trotzdem ich in etwa 4 Wochen nach Warschau kommen werde, wo ich meine mir seit fast anderthalb Jahren verlobte Braut holen werde, so darf ich doch vorderhand an weiter nichts denken, als, meinen stricten aerztlichen Vorschriften gemäss, sofort auf das Landhaus meines Vaters zu gehen und dort so ruhig als möglich den Sommer zu verleben. Dr. Kleinenberg u. Dr. Eisig verwalten während meiner Abwesenheit, wie auch schon jetzt während meiner Krankheit, die Zool. Station.—

Ihre freundl. Anfrage, betreffs Büchersendungen ecc. zu Gunsten der Station kann ich dahin beantworten, dass vor Allem die Royal Society,—wiederum durch Huxley's Vermittlung—am glänzendsten sich bewährt hat. Aber auch andre Akademieen,—Kopenhagen, Amsterdam, Neapel,

Wien, Berlin, Smithsonian Institution, haben viel geschenkt. Die Petersburger Akademie hat es zugesagt, aber bis jetzt habe ich noch nichts erhalten. Woran mir sehr viel liegt, ist, die Publicationen aller Russischen Universitäten soweit die Biologie davon betroffen wird, zu erhalten. Ich denke, allmälig dies zu erbitten. Gegenwärtig freilich habe ich alle Anstrengungen darauf zu concentriren, mehr Geldmittel zu beschaffen, um die Laboratorien mit Instrumenten und allerhand Bequemlichkeiten auszustatten.

Welchen Erfolg diese Bemühungen finden werden, das steht dahin, aber ich werde mit Ihrer Erlaubniss so frei sein Ihnen davon gelegentliche Mittheilung zu machen, da Sie doch einmal die Zool. Station unter Ihre Protection genommen haben.

Und so danke ich nochmals auf das Herzlichste für die vielfachen Beweise dieser ehrenden und fördernden Protection und wünsche von ganzem Herzen, dass Ihnen noch recht lange Gesundheit und Theilnahme an den Dingen dieser Welt möge zu Gebot stehen!

Den Grafen Tolstoi aber will ich regelrecht belagern, bis er auch den zweiten Tisch herausrückt,—und dazu sollen mir die jetzt hier anwesenden Russen behilflich sein.

In der Hoffnung, Ihre Verzeihung für die unverschuldete lange Verzögerung dieser Antwort zu erhalten bleibe ich

in innigster Verehrung
Ihr
Anton Dohrn

In 4 Wochen denke ich in Stettin im elterlichen Hause zu sein.

---

[20] Dohrn an von Baer
28.5./[9.6.1874], Petersburg, Telegramm

Dorpat Akademiker Carl Ernst von Baer.

Sind Sie dieser Tage zu sprechen—Rückantwort bezahlt.

Anton Dohrn

---

[21] von Baer an Dohrn
28.5./[9.6.1874], Dorpat, Telegrammtext

Ich bin zu Hause

Doctor K. von Baer

Dorpat 28 May

[22] Dohrn an von Baer
11.6.1874, St. Petersburg

M B
St. Petersburg. Fantanka 13.
Haus Feodoroff.
11. Juni 1874.

Excellenz!

Trotz aller Bemühungen, meine hiesigen Angelegenheiten rasch abzuwickeln, um im Stande zu sein, Ihnen meinen Besuch machen zu können und persönlich für Alles zu danken, womit Sie mich in meinen Bestrebungen unterstützt haben, sehe ich doch, dass ich diese Absicht aufgeben muss, und das thut mir sehr, sehr leid.

Es ist die Zoologische Station selbst, die mich zu dieser Entsagung zwingt,—und mein einziger Trost ist, dass ich dafür Erfolg habe.

Es wird Sie freuen zu hören, dass es mir gelungen ist, das Herz des Finanz-Ministers Reutern zu rühren, und ihn zu bewegen, auch einen Credit für einen zweiten Tisch zu eröffnen, gleichfalls auf 5 Jahre. Von Seiten des Grafen Tolstoy [*sic*] war das von vornherein beantragt worden, hatte aber bei Herrn Reutern Widerspruch erfahren, den ich glücklich beseitigt habe.

Aber l'Appetit vient en mangeant,—gegenwärtig bin ich noch auf den Beinen, um von der medizinischen Akademie einen dritten Tisch erwerben zu lassen. Von meiner Regierung bin ich an den Botschafter Prinz Reuss, und von diesem an den Kriegsminister empfohlen worden, bei dem ich demnächst eine Audienz haben werde. Hoffentlich gelingt der Plan.

Von hier gehe ich dann Anfang nächster Woche nach Stockholm und Kopenhagen in gleicher Angelegenheit, wennschon mit weniger Hoffnung auf Erfolg.—

Die englische Subscription, von der ich in meinem letzten Briefe schrieb, hat bis jetzt zwischen 8-900 Pfund Sterling eingetragen,—eine sehr grosse Hilfe! Weiterhin habe ich in Berlin versucht, mein Deficit zu verringern und hoffe in diesem October 5000 Thaler von der Preussischen und im nächsten October dieselbe Summe von der Deutschen Regierung zu erhalten. Sagt der Kriegsminister Milutine [*sic*][h] zu, so sind damit 14 Tische vermiethet und das repraesentirt eine Einnahme von 7000 Thalern,—damit nähert sich mein Ur-Plan der Vollendung: disponible Ueberschüsse zu gewinnen, mittelst welcher die Zoolog. Station in Stand gesetzt werden soll, selber jüngere Kraefte zu engagiren, um eine sorgfältige Fauna des Golfes ausarbeiten zu lassen,—und besonders die physikalisch-chemischen, geologischen u. botanischen Lebensbedin-

---

[h] Correct: Miljutin.

gungen dieser Fauna zu untersuchen, und gleichsam statistisch festzustellen. Ein weites Ziel,–aber doch keine Utopie.–

Darf ich Ihnen, Herr v. Baer, noch in aller Bescheidenheit eine Angelegenheit vortragen, die mich schon seit einiger Zeit interessirt hat, und von der ich in einer kleinen Schrift gesprochen habe, welche ich seiner Zeit glaube, auch Ihnen zugeschickt zu haben. Durch Beschäftigung mit den Anfaengen der Embryologie in ihrer historischen Entwicklung sind meinem Freunde Kleinenberg und mir, die gewaltigen Leistungen Caspar Friedrich Wolffs immer näher und näher getreten und wir haben Beide bedauert, dass die Petersburger Akademie noch nicht alle Schriften des grossen Mannes, die im Mss. vorhanden sein sollen, herausgegeben hat. Da Sie, Excellenz, früher diese Sache einmal in der Hand gehabt haben, wollte ich mir die Anfrage erlauben, ob hier vielleicht irgend ein Hinderniss vorliegt, dass [*sic*] sich vielleicht überwinden liesse. Leider sind die meisten der Herren Akademiker gegenwärtig schon auf Reisen, so dass ich meine Erkundigung direct bei Ihnen anbringen muss. Liesse sich vielleicht die Akademie bestimmen eine neue Gesammt-Ausgabe von Wolff's Schriften herauszugeben? Oder würde man eine solche Aufgabe vielleicht, falls keine grössere Kraft sich der Sache annähme, dem Dr. Kleinenberg,–der ja von Geburt Deutsch-Russe ist [–] und mir, dem Landsmanne Wolffs, anvertrauen? Wir würden Beide sehr stolz sein, eine solche Aufgabe der Pietät, die doch auch zugleich von so hoher wissenschaftlicher und historischer Bedeutung wäre, mit Aufbietung all unsrer Kraft durchzuführen. Aber freilich, zuerst müsste es überhaupt möglich sein.–

Herr v. Bradke, im Ministerium der Volksaufklärung, hat mich beauftragt, Ew. Excellenz seine besten Grüsse zu überbringen. Leider muss ich sie nun schriftlich mittheilen. Zugleich theilte er mir mit, dass dem Hrn. Dr. Rosenberg die Benutzung des ersten der beiden Russischen Tische freigestellt wäre, so dass dieser Herr sich sobald er will an Dr. Kleinenberg,–seinen Studienkameraden, wenden kann um alle Einzelheiten zu regeln. Sehr gern hätte ich Dr. Rosenberg noch persönlich gesprochen, das Fatum und General Milutine wollen es aber nicht.

Da ich höchstens noch bis Dinstag hier bleibe, so ist schwerlich Aussicht vorhanden, dass ich von Ihnen, Excellenz, eine wenn auch nur kurze Benachrichtigung empfinge,–sollte es doch möglich sein, so würde ich mich sehr freuen, besonders, wenn Sie hinzufügten, dass es Ihnen gut geht, und dass Ihre Gesundheit trotz des schlechten Wetters sich aufrecht erhält.

Mit diesem herzlichen Wunsche und der Bitte, meine wiederholten Anliegen freundlich aufnehmen zu wollen, bin ich

Ihr
treu und dankbar ergebner
Anton Dohrn

Bitte mich Herrn Dr. Rosenberg freundlichst empfehlen zu wollen.

[23] von Baer an Dohrn
25.7./6.8.1874, Dorpat

An Herrn *Dr. Anton Dohrn.*

Dorpat. $\frac{\text{25 Juli}}{\text{6 Aug}}$ 74

Hochgeehrter Herr Doctor!

Sehr leid hat es mir gethan, dass die Umstände es Ihnen nicht erlaubt haben, mich in Dorpat zu besuchen. Hätte ich nur gewusst, dass Sie nach St. Petersburg reisen würden, so hätte ich mich dahin begeben, da ich ohnehin daselbst noch einige Geschäfte abzumachen habe. Allein Ihre Ankunft hierselbst war mir schon vorläufig durch eine Dame angezeigt, und während ich auf Ihren Besuch hoffte, bekam ich das Telegramm aus St. Petersburg, das dieselbe Hoffnung nährte und einige Zeit später den Brief, der Ihre Abreise anzeigte, den letzteren Brief aber erst an demselben Tage, den Sie zur Abreise bestimmt hatten. Es war also auch keine Möglichkeit, darauf zu antworten. Später bin ich selbst aus Dorpat abwesend gewesen.

Es freut mich sehr, dass Ihre Wünsche in St. Petersburg in Erfüllung gegangen sind. Ich hätte Ihnen auch nichts besseres rathen können als sich unmittelbar an die betreffenden Behörden zu wenden, denn man ist in Russland durchaus zuvorkommender und rücksichtsvoller gegen Fremde als gegen Inländer.

Ich muss nun noch einem Wunsche entgegenkommen, mit dem Sie mich früher beehrt haben, indem Sie meine Photographie wünschten. Ich darf wohl annehmen, dass Sie eine Photographie im Formate der Visitenkarten meinten; aber auch dann noch blieb ich in dubio, ob ich Ihnen die letzte Photographie von mir zuschicken sollte oder eine frühere, da erstere mich eben nur als abgängig darstellt. Ein Abiturient vom Leben empfiehlt sich weniger als ein Abiturient von der Schule oder von der Universität. Ich wollte Ihnen also die Wahl lassen zwischen der letzten Photographie und einer vor fünfzehn Jahren in Frankfurt angefertigten. Bei persönlicher Zusammenkunft hätte sich das leicht machen lassen; jetzt kann ich mir nur dadurch helfen, dass ich Ihnen beide schicke und Ihnen die Wahl überlasse.

Ich nehme an, dass Sie noch in Stettin bei Ihrem Herrn Vater sich befinden, und schicke den Brief deswegen dahin.

Mit vollkommenster Hochachtung
Dr K E von Baer

[24] von Baer an Dohrn
26.7./7.8.1874, Dorpat

An Herrn Dr. Anton Dohrn.

Dorpat, 26 Juli / 7. Aug. 74.

Hochgeehrter Herr Doktor!

Kaum war gestern der Brief an Sie auf die Post geschickt, als mir beifiel, daß ich einen Punkt gar nicht beantwortet hatte, nämlich die Frage über die literarische Nachlassenschaft von Casp. Fried. Wolff. Ich bitte also dieses Blatt als eine Nachschrift zu dem früheren Briefe zu betrachten.

Es besteht, soviel ich weiß, nur ein von K. F. Wolff hinterlassenes Werk. Dieses Werk behandelt nur Doppel-Mißgeburten und zwar nur vom Menschen. Es ist sehr speciell und von einer großen Menge, sauber ausgeführter Zeichnungen begleitet. Ich hatte einmal versprochen, die Herausgabe dieses Nachlasses zu besorgen. Allein ich schreckte zurück vor den großen Kosten, welche diese Masse von Zeichnungen verursachen würden. Besonders schien mir der Gewinn der Wissenschaft gering im Verhältniß zu diesen Kosten, da die Mittel der Akademie leider sehr beschränkt sind. Die Hauptresultate der Untersuchungen über Doppelmißbildungen, z.B. daß die [?] gleichnamigen [?] zweier, wie man sagt verwachsener Individuen, zusammentreffen, sind längst gewonnen und haben gelehrt, daß man es hier nicht mit Verwachsungen, sondern mit Theilungen zu thun hat. Die Akademie hat gewiß unverantwortlich darin gehandelt, daß sie das Werk nicht früher herausgegeben hat; ob sie aber jetzt noch die Kosten darauf verwenden soll, dürfte sehr fraglich sein. Werden die Abbildungen im Auslande angefertigt, so werden sie nicht nur wohlfeiler, sondern auch besser ausgeführt. Daß Sie, geehrter Herr Doktor, und Ihr trefflicher Gehülfe, Dr. Kleinenberg, sehr geeignet sind, dieses Werk herauszugeben, versteht sich von selbst und wird von der Akademie gewiß nicht bezweifelt werden. Glauben Sie es herausgeben zu können ohne Beihülfe der Akademie, so wird letztere ein solches Anerbieten gewiß sehr freudig aufnehmen. Soll aber die Akademie die Kosten tragen, so ist mir ihre Entscheidung sehr zweifelhaft. Es werden der Akademie immer mehr Arbeiten angeboten, meistens mit zahlreichen Abbildungen, als sie an's Licht setzen kann. Worauf die geringen Mittel verwendet werden, hängt von der allgemeinen Konferenz, d. h. von der Zustimmung aller Akademiker ab.

Ganz ergebenst
Dr K E v Baer

[25] Dohrn/CAD an von Baer
10.8.1874, Stettin

Stettin. 10 August 1874.

Excellenz!

Sie haben mir durch Ihren Brief und die begleitenden Photographien eine ausserordentliche Freude bereitet. Aber wie sollte ich es über's Herz bringen, eine davon zurückzuschicken? Abiturienten sind wir ja Alle; und ich fürchte sehr, mein Lebens-Examen wird nicht so glänzend ausfallen, dass ich Anspruch auf die Achtziger erwürbe, besonders da sich der Locus minoris resistentiae längst bei mir eingestellt und durch die letzten Jahre sehr scharf accentuirt hat.

Meine Bewunderung dem Greise auszusprechen, der das grosse Problem gelöst hat, schön, gross und lange zu leben, werde ich nicht müde, und doppelt betrübt es mich, an Dorpat nothgedrungen vorbeigefahren zu sein, ohne diese Bewunderung in bescheidener Weise aussprechen gekonnt zu haben.

Nachtraeglich muss ich es noch mehr bedauern, als der Anlass meines laengeren Verweilens in St. Petersburg, nämlich der dritte, von der Medizinischen Akademie zu erwerbende Tisch, in's Wasser, leider der Newa nicht des Golfs v. Neapel, gefallen ist. Ich höre wenigstens, dass trotz entschiedener Befürwortung des Kriegsministers, – dessen Brief an den Fürsten v. Reuss ich habe, – die Akademie den Tisch abgelehnt haben soll. Ich will noch einmal directe Schritte beim General Milutine thun, bei dem ich sehr gut introducirt bin, – aber ich fürchte mit Intriguen zu schaffen zu haben, oder mit Empfindlichkeiten, deren Beseitigung schwer, wenn nicht unmöglich sein wird.

Gegenwärtig habe ich die Freude, Ihre Studien über die Flüsse Russlands zu lesen; auch habe ich wiederholt den Protest gegen die Teleophobie, und, wie ich aufrichtig sagen darf, mit viel innerer Genugthuung gelesen. Ich gestehe Ihnen ganz offen, Excellenz, dass ich ein entschiedener Anhaenger der Darwin'schen Theorie bin, und dass ich überzeugt bin, die Ansprüche der edleren Naturen lassen sich durch Vertiefung und Ausbau des grossen Gedankens befriedigen. Das Huronengeheul, das von gewisser Seite ausgeht, degoutirt mich im Innersten, – aber leider findet es in unserm brutalen Zeitalter zuviel Boden, – und zu viel Auflagen. Just weil ich Herz und Seele dem Gedankenbau Darwin's verschrieben habe, kränkt es mich doppelt, diesem wüsten Treiben zusehen zu müssen. Gegenwärtig sind mir aber aus 100 Gründen die Haende gebunden, und ich darf nicht, trotz innigsten Wünschens, als Darwinianer reinen Wasser's gegen diese Darwinianer *un*reinsten Wassers vom Leder ziehen. Die Art wie David Strauss diesem Unfug in die Haende gearbeitet hat, fehlte grade noch, um den Veitstanz vollkommen zu machen, und es ist gar nicht abzusehen, wohin die wilde Jagd noch gehen mag. Um so begieriger bin ich auf Ihr verheissenes Werk.

Mit der Bitte, die einliegende, nach Aussage der Meinigen viel bessere Photographie an die Stelle der früher gesandten setzen zu wollen und mir die Genugthuung zu gestatten, *zwei* Portraits von Carl Ernst v. Baer in mein Album zu stecken, grüsse und empfehle ich mich auf das Beste und bitte um Fortdauer Ihrer Theilnahme an meinem seiltänzerischen Unternehmen in Neapel

In Verehrung und Anhaenglichkeit
Anton Dohrn

[CAD:]
Der Commissionarius, dem aus vorliegendem Grunde die BriefbeförderungsCommission leicht genug wurde, benutzt gern den willkommenen Anlass, seinen und des eben zur Eisenbahn abmarschirten Dr. Heinrich's herzlichsten Gruss beizufügen. Da ich gerade mir den Nebenscherz erlaube, ein Stück von Calderon zu übersetzen, so wird der spanische Wunsch passen

Viva V[uestra]m[erce]d mil años
Ihr C. A. Dohrn.

Auf meinem Tusculum
am 11 August 1874.

---

[26] von Baer an Dohrn
[eingegangen: 10. Januar 1875, Dorpat]

Hochgeehrter Herr Doctor!

Um Ihnen das neueste Beispiel einer sichtbaren Transformation vorzuführen, überschicke ich die beifolgende Photographie. Es ist das neueste Bild, des Ihnen bekannten Dr. K. E. v. Baer, der sich seit einigen Jahren einen Bart hat wachsen lassen, und den man deßwegen genöthigt hat, noch einmal einem Photographen zu sitzen. Der Mann sieht sehr verdrießlich aus, ohne Zweifel, weil er seine Augen nicht mehr brauchen kann.

Soeben habe ich mit der Zeitschrift für wissenschaftliche Zoologie das Verzeichniß der Bibliothek Ihrer Anstalt erhalten. Diese Bibliothek ist ja schon sehr reich. Man könnte ein Jahr hindurch in Neapel sitzen, nur um in der Zoologischen Literatur sich zu orientiren, wenn man das Glück hat, noch seine Augen gebrauchen zu können.

Ich habe in der letzten Zeit einen Aufsatz über den Darwinismus verfaßt, der für den zweiten Band meiner Reden bestimmt ist, wenn der Verleger ihn drucken will. Ich bin dabei halber Darwinist, oder vielmehr Transformist geworden; aber ein voller Anhänger kann ich eben so wenig sein, als ein voller Gegner. Ich bin dabei auf der alten Sandbank sitzen geblieben, welche die Ueberschrift führt: Nescimus. Eben deßhalb habe ich aber nichts dagegen, wenn Andere, welche mehr Hoffnung haben, vom Kap der guten Hoffnung aus ihre Untersuchungen anstellen. Wenn sie den Südpol auch nicht erreichen sollten, können sie ihm doch näher kommen. Das aber lasse ich mir nicht nehmen, daß alles Werden nach Zielen strebt, denn ohne Ziel kann ich mir kein Werden denken. Vielleicht geht von Neapel der Mann aus, der die Mittel des Werdens erkennt, von dem ich auf der letzten Seite der Vorrede meiner Entwickelungsgeschichte gesprochen habe.

Mit dem herzlichsten Glückwunsche zum Neuen Jahr, Ihr ergebenster

Dr K E v Baer

---

[27] Dohrn an von Baer
17.1.1875, Neapel

Napoli. Palazzo Torlonia. 17. 1. 75.

Excellenz!

Eben hatte ich die in der Abschrift beifolgende Einleitung in meine kleine Schrift "Der Ursprung der Wirbelthiere und das Princip des Functionswechsels'" beendigt, als ich Ihren Brief und die praechtige Photographie des so hoch verehrten Greises erhielt, der weder verdriesslich aussieht, noch auch verdriesslich zu sein scheint. Wollen auch die Augen nicht mehr von aussen Licht aufnehmen, innen ist es doch noch viel heller, als bei vielen modernen Propheten, die mit natürlichen und künstlichen Linsen ausgestattet sind, und an keine andern als ihre eignen Offenbarungen glauben.

Was werden Sie nun aber zu meiner Schrift sagen? Werden Sie erlauben, dass die Einleitung mit Ihrem Namen und an Sie gerichtet so abgedruckt wird, wie ich sie Ihnen schicke? Ich hoffe es um so mehr, als ich ausdrücklich gesucht habe, meine Meinungen eher als Ihnen entgegenstehende zu characterisiren, damit Ihnen nicht etwa nachgesagt werden könne, Sie hätten mich zu einer so revolutionären Arbeit angetrieben. Nur darin wollte ich mich zu Ihnen, nicht zu den orthodoxen Darwinianern zählen, dass ich noch lange nicht zugebe, mit der Natürlichen Züchtung sei der Kreis der Principien erschöpft, den uns die Entwicklung des Lebens aufzusuchen nöthigt; dass darum auch den metaphysischen Bedürfnissen so lange ihr Recht nicht abgesprochen werden dürfe, als die physischen Erklärungen so wenig zureichend sind.

Ich verstehe, oder glaube zu verstehen, was Sie verlangen, wenn Sie sagen, Sie können Sich ein Werden ohne Ziel überhaupt nicht vorstellen. Ich habe versucht, dies Element des Zielstrebigen in den Begriff der Vervollkommnungsfähigkeit hineinzudenken, und darin sowohl das teleologische wie das causale Element gleichmässig und einheitlich verbunden zu gestalten. Vorläufig enthalte ich mich jedes Versuches einer Definition dieses Begriffs und bin auch überzeugt, dass eine hinreichende Definition vorderhand fast unmöglich ist. Aber das Bedürfniss sollte doch einmal ausgesprochen werden, und, als Darwinianer ohne Furcht und Tadel, hielt ich mich für berechtigt, das zu thun.

Darin und in dem Hinweis auf die schrankenlose Degeneration, die gegenwärtig und immer Statt [*sic*] gefunden hat, erblicke ich die dauernden Elemente meiner Arbeit. Ob Amphioxus ein ursprünglicher oder ein degenerirter Fisch ist, ob die Ascidien so oder so zu verstehen sind, das mag der grossen Masse der Naturforscher vorläufig die Hauptsache sein, und sie mögen sich darüber gegenseitig die Haare ausraufen,—aber wenn es mir gelungen sein sollte, mit der Definition des Functionswechsels eine Wahrheit gefunden zu haben, die weitere Wahrheit nothwendigerweise produciren muss, dann will ich gern in vielen Einzelheiten Unrecht haben.—

Der Druck der kleinen Schrift, die bei Engelmann erscheint, soll in 3 Wochen fertig sein. Ich schicke Ihnen die Aushängebogen zu. Sie enthält drei Abschnitte: 1, Genealogischer Zusammenhang der Anneliden u. Wirbelthiere. 2, Genealog. Zusammenhang der Wirbelthiere und Ascidien. 3. Das Princip d. Functionswechsels. Die letzte Seite enthält einen Blick in die Unendlichkeit, der mir wahrscheinlich sehr verübelt werden wird,—ist mir doch schon früher jede Befähigung zur Naturforschung abgesprochen worden, weil—ich zu viel Musik triebe! αὐτὸς ἔφα in Jena.—

Dass Sie so freundlich sind, der Zool. Station wieder in so anerkennender Weise zu gedenken, freut mich doppelt, da gleichzeitig mit Ihrem Briefe eine andre Anerkennung kam mit der Meldung, der Kaiser habe am 2ten Januar ein Decret unterzeichnet, das der Station zum zweiten Male 10,000 [Thaler] vom Reich zuwendet. Damit wird ein grosser Theil Schulden wegbezahlt. Auch sind die sichern Einnahmen grösser geworden, ja ich hoffe, bald einen vollkommnen Gleichgewichtszustand zwischen Credit u. Debet erreicht zu haben.

Doch über Alles Das hoffe ich Ihnen im Mai dieses Jahres mündliche Mittheilung machen zu können, denn ich werde über Wien, Pest, Moskau nach Petersburg gehen und dann nicht zum zweiten Male an Dorpat vorbeigehen. Ich will sehen, ob ich bis 20 oder gar mehr Tische mit festen Contracten vermiethen kann,—dann erst wird mein Unternehmen feststehen.—

Darf ich Sie nun fragen, ob ich die Erlaubniss erhalte, die übersandte Einleitung abzudrucken? Ich erwarte diese Erlaubniss natürlich erst, wenn Sie die Schrift selbst gelesen haben werden, die Ihnen in Aushaengebogen von Leipzig zugesandt werden soll. Zugleich sende ich an Engelmann das Mss. der Einleitung, und würde dann bitten, dass Sie so freundlich wären, zwei Zeilen an ihn zu richten: drucken Sie oder drucken Sie nicht die Einleitung.

Wie Sie auch entscheiden, es bleibt doch wahr, dass ich meine Schrift innerlich an Sie gerichtet habe. Verzeihen Sie die Vermessenheit.

Und nun wünsche ich vor Allem Fortdauer der Geistesfrische und Heiterkeit, die aus Ihrem Briefe spricht, und dass uns bald durch den Verleger zugänglich werde, was Sie über den Darwinismus zu sagen haben.

Mit herzlichem nochmals wiederholtem Dank und der Versicherung unveraendlicher Anhaenglichkeit

Ihr
Anton Dohrn

---

[28] Dohrn an von Baer
6.2.1875, [Neapel]

Excellenz!
Ich eile ein Versehen des Hrn. Buchhaendler Engelmann, soweit ich kann, gutzumachen und Ihnen zu schreiben, dass Hr. Engelmann von mir ersucht ward, die Widmung an Sie *nicht eher* im Druck zu setzen, *als bis Sie Ihre Zustimmung darüber ausgesprochen hätten,* und die Erlaubniss gegeben, sie zu drucken.

Ich möchte nun recht von Herzen bitten, dass falls Sie das geringste Bedenken dagegen haben, Sie es mir rückhaltlos zu erkennen geben,— denn so sehr ich wünschte, dass die kleine Schrift Ihren Beifall faende, so sehr bin ich doch davon durchdrungen, dass solche Freiheiten, wie sie in Widmungen und andern rein persönlichen Angelegenheiten liegen, nur in der rücksichtsvollsten Weise ausgeübt werden dürfen. Die gute Sitte und der Anstand werden jetzt oft so misshandelt, dass ich um keinen Preis in Ihren Augen als Einer dastehen möchte, der an dieser Gesinnungsart theil hätte. An der Meinung der Masse liegt mir gar nichts, aber an Ihrer Meinung Alles.

Sie sehen, dieser Brief ist von dem Egoismus dictirt, von Ihnen nicht getadelt zu werden,—und dass dies eine aufrichtige Aeusserung meiner Empfindungen ist, werden Sie mir glauben.

In grosser Eile, um nicht zu spät zu kommen

Ihr
treu ergebner
Anton Dohrn

Pal. Torlonia. 6. Februar 1875.

[29] Dohrn an von Baer
20.2.1875, Neapel

Napoli. Pal. Torlonia. 20. 2. 75.

Excellenz!

Trotzdem es unbescheiden aussieht, wenn ich schon wieder ohne erst Ihre Antwort abzuwarten, an Sie schreibe, so muss ich die Verantwortlichkeit hiefür auf den 29sten Februar 1792 wälzen, an welchem Tage, —laut Carus & Engelmann's Bibliotheca Zoologica—der Verfasser der "Entwicklungsgeschichte der Thiere" geboren ward.

Ihm zu seinem 83sten Geburtstage die schuldige Ehrfurcht auszudrücken, ist der Auftrag dieser wenigen Zeilen, nach dessen Erledigung der Schreiber dieses auch sofort wieder von der Bühne verschwindet, trotzdem es aber nicht versäumt, sich zu unterschreiben als

Ihr
herzlich und treu ergebner
Anton Dohrn

[30] von Baer an Dohrn
8./20.2.1875, Dorpat

Dorpat d. 8ten/ 20. Februar 1875

Hochgeehrter Herr Doktor!

Ich habe nicht nur Ihre geehrte Zuschrift erhalten, sondern auch von Herrn Engelmann den Abdruck der Widmung so wie des ersten und dritten Bogens des Textes. Ich kann für Ihre Zuschrift an mich nur danken,—da sie über Verdienst ehrend ist. Ich habe also dankbar meine Zustimmung gegeben. Natürlich habe ich auch etwas in die Abhandlung selbst hineingeguckt. Dabei hat es, wie Sie voraussetzen, an einigem Kopfschütteln nicht gefehlt. Daß die Anneliden sich umkehren, um zu Wirbelthieren zu werden,—war mir schon etwas geläufig und daher weniger anstößig. Aber daß die Extremitäten, und namentlich die der Wirbelthiere, aus Athmungsorganen geworden seien, das ist mir sehr anstößig. Sollen denn die Thiere gar nichts mehr nach eigenem Bedürfniß bilden können, sondern nur Ererbtes moduliren? Ist die freie Beweglichkeit in der Natur des Thieres begründet, und werden die Mittel dazu im Wasser in verschiedenster Weise gegeben, so scheint es, daß für die Thiere, die ihre Nahrung auf festem Boden finden sollen, entweder Flügel od[.] Hebelreifen da sein müssen.

Aber ich sehe wol, daß die neue Art der Weltanschauung eine ganz andere ist als die, an welche ich mich gewöhnt habe. Deßwegen möchte ich mich gar nicht mehr im Areopag der Naturforscher vernehmen lassen, wenn ich nur vorher einen Aufsatz über den Darwinismus los werde. Ich weiß nicht, ob ich Ihnen geschrieben habe, daß ich seit längerer Zeit einen solchen Aufsatz dictando vorbereitet habe.

Nun aber liegt er und mein Verleger kommt zu keinem Entschlusse, wie er es machen soll, daß die Korrektur gehörig besorgt werde. Doch das sind curae minores, die nicht über Dorpat hinauszukommen haben. Diese Abhandlung über den Darwinismus entscheidet sich weder für noch gegen denselben, sondern wägt Bedenken und Gegenbedenken gegeneinander ab. Ich kann nicht umhin, die Transmutation in hohem Grade wahrscheinlich zu finden, kann aber die Darwin'sche Selektions-Hypothese nicht für genügend erklären, sondern glaubte die Transmutation für eine Entwicklungsform erklären zu müssen. Wenn ich den Herbst dieses Jahres noch erleben sollte, so würde es mich sehr freuen Sie persönlich kennen zu lernen, ja, ich habe in der Mitte des Winters, zu welcher Zeit ich immer sehr reiselustig bin, schon eine Reise bis Neapel geplant, d.h. nicht für den Winter, sondern für den Sommer und Herbst. Aber ich fürchte, man wird mich nicht reisen lassen, da die Meinigen mich immer noch gebrechlicher finden als ich selbst. Nun, jetzt sind wir noch von tiefem Schnee umgeben, was die Zukunft bringt, wird sich erst zeigen, wenn der Schnee fort ist, der in diesem Jahre sehr permanent scheint.

Mit herzlicher Ergebenheit
Der Ihrige
Dr K E v Baer

---

[31] Dohrn an von Baer
1.4.1875, Neapel

Napoli. 1. April. 1875.
Palazzo Torlonia.

Excellenz!

Jetzt wird wohl meine kleine Schrift in Ihren Haenden sein, ich darf Ihnen also noch einmal meinen Dank dafür aussprechen, dass Sie mir gestattet haben, Ihren Namen derselben voranzudrucken.

Ich wünschte freilich, es wäre Ihnen möglich gewesen, wenigstens den Weg gutzuheissen, auf dem ich die genealogischen Resultate zu erreichen hoffe, und das Princip anzuerkennen, das ich gefunden zu haben glaube. Die Ergebnisse meines Nachdenkens weichen ja freilich sehr weit vom heut Angenommenen ab,–aber ist das an sich schon ein Fehler? Wir erleben jeden Tag ein neues "Heute", und sollte ich für das heutige Heute einen Fortschritt angebahnt haben,–wie lange wird er diesen Namen verdienen, gegenüber den ungeheuren Gebieten, auf denen uns kein Wegweiser zu Gebote [steht]. Ich hoffe, der "Functionswechsel" werde für die nächsten Jahre freilich eine grosse Hilfe zur Aufsuchung und Lösung einer Reihe morphologischer u. physiologischer Probleme sein,—aber er ist eben auch nur ein Modalitätsbegriff,—und das Leben bleibt uns schliesslich ebenso unverständlich.

Doch Alles Das wird Ihnen meine Schrift nicht sympathischer machen, und ich muss mich trösten mit dem Bewusstsein, dass ich trotz Ihrer abweichenden Meinung nicht aufzuhören brauche, in Ihnen und in Darwin die beiden Ecksteine des gegenwärtigen Gebäudes der Biologie zu sehen, und darauf zu hoffen, daß eine weitere Entwicklung unsres Erkennens auch die Unterschiede harmonisch ausgleichen werde, die noch zwischen den von Ihnen angebahnten Denkweisen und der Theorie der Natürlichen Züchtung bestehen.

Mich selbst, Excellenz, halten Sie aber immer als einen Ihrer grössten Verehrer, der sich aufrichtig bemüht, Ihnen auch [*sic*] denjenigen Pfaden des Denkens zu folgen, die von Anfang an von der eignen Grundlage verschieden zu sein scheinen, und der wenigstens verstehen will, wenn er auch nicht das Verstandene theilen kann.

Und so wünsche ich, daß Sie bald ein bessres Medium aufsuchen möchten zum Leben, als den Schnee u. Eis Russlands, und erbiete mich zu jedem Reise-Marschalls-Dienst, falls Sie wirklich Ihre Absicht ausführten und uns Kindern des Südens einen Besuch machten.

Mit der Versicherung treuester Anhaenglichkeit

Ew. Excellenz
sehr ergebner
Anton Dohrn

---

[32] von Baer an Dohrn
11./23.6.1875, Dorpat

Dorpat, d. 11. resp. 23. Juni 1874. (5)[i]

Hochgeehrter Herr Doctor!

Es thut mir herzlich leid, daß meine kleinen Bemerkungen über Ihre wichtige Schrift Sie etwas aufgeregt zu haben scheinen; auch habe ich wohl Unrecht gethan, meine abweichende Meinung von der Entstehung der Extremitäten der Wirbelthiere zu derb hervorgehoben zu haben und von dem, was in Ihrer Schrift mir sympathisch gewesen ist, zu wenig gesagt zu haben. Vor allen Dingen hat es für mich viel mehr Ansprechendes, die Wirbelthiere von den Ringelwürmern herzuleiten, die doch in Antimeren getheilt sind, als von den ganz ungegliederten sackförmi-

---

[i] The last number (4) has been underlined by someone else and corrected to "(5)".

gen Ascidien, deren Muskelhaut auch nur ein Sack ist. Für diese Ansicht sehe ich gar keinen andern Grund, als die Beobachtungen über die Entwicklung von Amphioxus und Ascidien, von denen die erstern nicht einmal contrahirt scheinen. Eine Ableitung von den Rundwürmern würde in die Augen springen, wenn man eine Umkehrung von Bauch und Rücken nachweisen könnte. Sie haben mit dem fraglichen Durchbruch des Maules sicher die Achilles Ferse dieser Ansicht getroffen; nun kommt es darauf an, ob ein solcher Durchbruch erwiesen oder wenigstens wahrscheinlich gemacht werden kann. Mir fiel sogleich aus dem veralteten Vorrath meiner Erinnerungen *Siphonostoma diploehaites* [*sic*] von Otto ein, die im zehnten Band der Denkschriften der Leopold. Carol. Academie vorkommen. Ich weiß garnicht, was später aus dem Berichte der hintereinander liegenden Mundöffnungen geworden ist, d.h. wie sie vernünftig gedeutet werden. In dem Handbuche von Claus finde ich keine Belehrung darüber. In andern Büchern auch nichts weiter, als Zurückführung auf andere genera, die ich jetzt nicht prüfen kann. Sie werden das als Anwohner des Mittelmeeres u[.] Kenner der niedern Thiere viel besser wissen u[.] ich muß annehmen, daß Sie auf dieser Fährte keinen Gewinn erwarten, daß Sie nicht auf dieselbe eingehen.–

Das Princip der Degeneration hat mir auch sehr zugesagt, denn ist überhaupt eine Umwandlung möglich, warum nicht auch ein Rückschritt, den wir überdies bei den Schmarotzerkrebsen so kolossal noch im Lebenslauf des einzelnen Individuums sehen. Auch gegen den Functionswechsel habe ich gewiß nichts; ist er doch in geringerem Grade in luftathmenden Wasserschnecken von Siebold empirisch erwiesen.–

–Aber darin sind unsere Ansichten verschieden, daß ich gar keine Nöthigung fühle, die Wirbelthiere von einer andern größern Gruppe herzuleiten. Amphioxusartige weiche Wirbelthiere können sehr lange bestanden haben, ohne deutliche Spuren hinterlassen zu haben. Ja ich vermute sie unter manchen sogen. *Graptolithen.* Ganz weichen wir aber darin ab, daß Sie anzunehmen scheinen, wie die meisten Darwinisten, ein Thier könne keinen Theil besitzen, den es nicht ererbt habe, daher auch unsere Arme und Beine bei den Rundwürmern Kiemenstämmchen gewesen sein müssen, ich dagegen die Ueberzeugung habe, es müsse jede regelrechte Entwicklung eines Organismus, (also mit Ausnahme der der Miß-rathenen oder Miß-geburten) darauf gerichtet sein, eine auf unserer Erde mögliche Existenz zu schaffen. Soll also, nach meiner Meinung, ein Organismus aufgebaut werden, der seine Nahrung auf dem festen Erdkörper zu suchen hat, so ist damit auch verbunden, daß er gelenkige Füße erhält, u[.] zwar ganz nach den Gesetzen der Mechanik. Darwin wird freilich sagen, wenn ein Grasfresser oder auch ein Landraubthier ohne Beine geboren wird, so muß eine Creatur dieser Art im Kampf um das Dasein zu Grunde gehen. Ich aber behaupte, ein solches Gebilde ist gar nicht anders möglich als durch die entschiedenste Theratologie.

–Mein Aufsatz über den Darwinismus ist allerdings abgegangen, aber ich kann nicht sagen, daß ich mit ihm zufrieden wäre, denn ich kann die Mutabilität nicht leugnen, für die ich mich ja auch vor Darwin ausgesprochen habe. Ich kann aber für Darwin's Erklärung derselben d.h. für die Selektionslehre mich garnicht erklären, sondern trete dagegen auf. Sie geht von dem Gedanken oder dem Gefühle aus, daß in der Natur keine Vernunft wirken könne, ich dagegen kann mich von der alten Ansicht nicht losreißen, daß Vernunft oder Zielstrebigkeit in ihr ist, mag diese immanent sein, oder transscendent [*sic*]. Wenn Jemand möglichst richtige Kreise gezeichnet wünscht und unter Millionen Versuchen, die mit freier Hand gezeichnet sind, die besten auswählt, so werden sie doch nicht so richtig sein, als wenn mit einem Zirkel d.h. mit einer zielstrebigen Nothwendigkeit eine Bogenlinie in immer gleicher Entfernung um einen Punkt geführt wird. Die zweite Methode wird also wohl besser zum Ziele führen, als die Zuchtwahl, welche eine unmeßbare Anzahl von Versuchen verwerfen muß.–

–Ich hatte in einem viel früheren Briefe die Hoffnung geäußert, nach Süden bis in Italien zu reisen, um gleichsam von der Welt Abschied zu nehmen. Da ich endlich mein Manuscript losgeworden bin, fühlte ich mich als freier Mann. Leider scheint das anders werden zu sollen. Ich habe an einem schon früher schadhaften Unterschenkel ein Geschwür bekommen, das nur sehr spät oder gar nicht heilen zu wollen scheint. Vor der Heilung kann ich aber wohl nicht reisen. Erfolgt dieselbe nach einem oder zwei Monaten, so würde ich im Herbst wohl noch irgend eine Reise über die Grenze antreten. Ich wünsche es zwar, bis Neapel zu kommen, aber ich darf es kaum hoffen. Jedenfalls dürfen Sie sich dadurch auf keine Weise abhalten lassen, denn ich könnte ja ihre [*sic*] Anstalt doch noch sehen u. würde H...[?] Kleinberg [*sic*], den ich zu grüßen bitte, vorfinden. Ich fürchte sehr, daß ich durch meine frühere Anzeige Sie von wichtigen Unternehmungen abgehalten habe. Ich bitte daher dringend, auf diese meine letzte mitgetheilte Hoffnung ja nicht Rücksicht zu nehmen; es ist ja sehr möglich, daß ich nur noch die große Reise in's Schattenland zu machen habe. Sie hier zu sehen, würde mich natürlich sehr erfreuen, allein in den jetzigen Sommermonaten ist in St. Petersburg gar kein Geschäft zu machen.

–Wie es auch kommen mag, wollen wir über den Darwinismus usw. nicht streiten, jedenfalls haben Sie mit der Gründung einer Anstalt, in der man den Entwickelungsgang sorgfältig studiren kann, indem die Embryonen sich möglichst conserviren lassen, einen Fortschritt eingeleitet, der den Speculationen von Darwin doch wohl Zaum u[.] Zügel anlegen wird, wenn diesem Zaum u[.] Zügel nothwendig sind. Finden sich die Zügel in Neapel nicht, so sind sie vielleicht gar nicht nothwendig.

Mit unveränderter Ergebenheit und Hochachtung

Ihr
unveränderlicher
Dr. K E. v. Baer

[33] Dohrn an von Baer
8.7.1875, Hökendorf

Hökendorf bei Alt-Damm.
Stettin. 8. Juli 1875.

Excellenz!

Mit Nichts hätte ich mehr und freudiger in meinem Vaterhause überrascht werden können, als mit dem langen und herzlichen Briefe, den Sie mir gesandt haben.

Es ist ganz wahr, dass es mir leid that, mit meiner Hypothese über den Ursprung der Wirbelthiere bei Ihnen keine Gnade gefunden zu haben. Ich glaubte, Sie würden den grossen Griff, den diese Hypothese über Vorwärts und Rückwärts thut, mit Sympathie auffassen, und in dem von mir postulirten Begriff der Vervollkommnungsfähigkeit eine Art Versöhnung zwischen Zielstrebigkeit, d. h. teleologischer Nothwendigkeit und causaler erblicken. Und ich gestehe Ihnen offen, ich wünschte sehr, Ihnen mündlich meinen Gedanken-Hinterhalt vorlegen zu können, nicht nur den speculativen, sondern auch den factischen, der in einer nächsten Schrift in einer morphologisch-genealogischen Erklärung unsrer Sinnes-Organe, zunächst Auge, Ohr und Nase, meiner Hypothese ein solches Uebergewicht oder Erklärungsfähigkeit sichern wird, dass sie all und jede Concurrenz mit Ruhe aus dem Felde schlagen wird. Aber einstweilen bin ich noch Sklave der Zoologischen Station und muss Frohndienste [*sic*] leisten, d. h. in der weiten Welt hin und her fahren, und die Mittel schaffen, um das so glücklich und erfolgreich Begonnene auch gegen Wechselfälle des Glücks in der Zukunft zu sichern. Dazu gehört zunächst, dass mir meine Landsleute in Anerkennung des nationalen Elements, das in dem Unternehmen steckt 10,000 Thaler schenken, um ein kleines Dampfschiff und verschiedene Einrichtungen in dem Gebäude der Station selbst zu beschaffen, dann aber muss ich die Zahl der fest vermietheten Tische bis auf 24 steigern, damit die Betriebskosten dadurch gedeckt, (wenigstens nothdürftig, da sie 13,000 Thaler betragen) und die Einnahmen des Aquariums zur Schuldentilgung und zur Ansammlung eines Reservefonds verwandt werden können. Ich hoffe im Jahre 1877 dies Ziel erreicht zu haben, Deutschland mit 10, Russland und Italien mit je 4, Oestreich-Ungarn mit 2 und die übrigen europaeischen oder ausser-europaeischen Staaten in wechselnden Vertraegen für die übrig bleibenden Tische engagiren zu können. Dann erst werde ich "Exegi" rufen. Italien hat bereits durch die offizielle Stimme seines Unterrichtsministers und durch das Gutachten seiner sämmtlichen Universitäten die 4 Tische für das nächste Jahr zugesagt, und wird den bezüglichen Vertrag auf zunächst 5 Jahre schliessen; die einzelnen deutschen Regierungen werden im nächsten Jahre im Ganzen 10 Tische inne haben und ich hoffe, es werde meinen diplomatischen Künsten gelingen, diese sämmtlichen 10 Tische auf das Reich übergehen zu lassen, die Einzel-Staaten aber zu vermögen, aus ihren Fonds Stipendien für die nach Neapel gehenden Forscher zu stiften. Da ich mit dem Herrn Finanz-Minister v. Reutern in persönlich nähere Berührung gekommen bin, und sein persönliches Interesse an der Zoologischen Station geweckt ist, so hoffe ich, es werde nicht schwer sein, von ihm das Geld für 4 Tische zu erhalten, falls Hr. v. Bradke und Minister Tolstoi die Ueberzeugung gewinnen, dass Russland mehr als genug Biologen hat, um diese Zahl von Tischen besetzt zu halten.

In Wien und Pesth war ich vor 8 Tagen und hoffe mit gewünschtem Erfolge, jetzt treibt mich die Subscription in Deutschland herum, und die Sorge für noch mehr Tische nach Kopenhagen und Stockholm, sowie nach England.

Da haben Sie, Excellenz, in Kürze den ganzen Schlachtplan. Ein offnes Sendschreiben, das ich an Prof. v. Siebold gerichtet habe, wird Ihnen bald allerhand Interiora der Station vorführen, die hoffentlich mit den mündlichen Berichten des Dr. Rosenberg doch ein Bild von der ganzen Sache gewähren können.

Dass Sie aber schon sollten die Koffer gepackt halten für die Reise zu den Schatten, das glaube ich Ihnen nicht. Mir haben Nachrichten ganz andre Mähr gebracht, und ich bin überzeugt, dass Sie den Marsch in die Neunzig mit sicherm Schritte fortsetzen. Ich rechne mit Gewissheit drauf, Ihnen noch die in Thatsachen gemünzten Beweise für die Abstammung der Fische von den Anneliden vorlegen zu können, freilich bin ich eben erst mit dem Fördern des edlen Metalles beschäftigt, das seiner Prägung in den nächsten 2 Jahren harrt. Aber die Prägstöcke sind fix und fertig,—des Gedankens Blässe kränkelt mich nicht mehr an. Bis dahin lass ich mich ruhig Phantasten, kritiklos, unwissend, und wie die Schmeicheleien alle heissen, welche liebe Collegen in Bereitschaft halten, nennen,—ich kenne all die guten Epitheta, sie bildeten auch die Allee, durch welche ich zur Zoologischen Station fortschritt,—es hat weiter nichts zu bedeuten.

Ihnen aber, Excellenz, drücke ich in treuer, dankbarer Verehrung die Hand und hoffe Sie in Neapel oder in Dorpat zu sehen, ehe wir ein Jahr aelter geworden sind.

Mein Vater und mein Bruder Heinrich tragen mir Beide die wärmsten Wünsche für Sie auf, und ich bitte freundlichst, mich den Herren Fachgenossen in Dorpat empfehlen zu wollen, besonders auch mit freundlichem Grusse Herrn Dr. Rosenberg, dessen Gegenwart in der Zool. Station mir eine grosse Freude, aber auch insofern ein Kummer war, als noch lange nicht Alles so geartet war, um einen so vortrefflichen Arbeiter zum vollen Ertrag seiner Mühen kommen zu lassen.

Ich werde nicht verfehlen von den weiteren Schritten zu Gunsten der Neapolitanischen Station zu melden, wie ich auch leider mitzutheilen habe, dass die Schöpfung von Agassiz, die Anderson School for Natural History zu Grunde geht, da weder oeffentliche noch private Mittel mehr zu beschaffen sind. Diessmal hoffentlich zeigt Europa America, wie man's zu machen hat.

Mit nochmaligem herzlichen Gruss und Dank

Ihr treuer
Anton Dohrn

---

[34] Dohrn an von Baer
18.8.1875, Stettin

Stettin. 18. August 1875.

Excellenz!

Diessmal kommt der Repraesentant der Zoologischen Station mit einem gar gewaltigen Anliegen an Sie.

Wie ich glaube, Ihnen schon früher geschrieben zu haben, bin ich seit Jahr und Tag damit beschäftigt, für die Station in Deutschland durch eine Subscription weitere Geldmittel flüssig zu machen. Das Ding stösst leider auf mehr Hindernisse als mir lieb und der Station zuträglich ist. Die allgemeine Handelskrise wird überall als Rechtfertigung vorgeschoben, dass selbst sehr reiche Leute entweder gar Nichts oder zu geringfügige Summen zusagen; und ich verbrauche unverhältnissmässig viel Kraft und Zeit zur Beschaffung unzureichender Mittel.

Eben von England zurückgekehrt habe ich mich überzeugt, dass ein kleines Dampfschiff, wie es für die Station nach vielen Seiten hin ausserordentliche Vortheile bieten würde, nicht unter 8000 [Thaler] zu beschaffen ist. Andrerseits bedarf ich zur Bestreitung der Kosten für die Anschaffung einer Dampfluftpumpe 800 [Thaler], für Neukauf von dicken Glasscheiben für das Aquarium ebenfalls 800 [Thaler]; die Einrichtung einer Sammlung (Beschaffung von Schraenken, Alcohol, Glaesern ecc.) einige Neubauten im Keller des Gebäudes zur Herstellung eines Kohlen-Magazines, einer kleinen Werkstätte meines Mechanikers und eines Speicher-Raumes für conservirte Thiere, sowie die Aufstellung und Ausrüstung von 6 neuen Arbeitstischen normiren sich auf mehrere Tausend Thaler, – kurz Alles in Allem würden mit 12–15 000 [Thaler], die mir vom Deutschen Publicum zur Verwendung übergeben würden, grade die Bedürfnisse des jungen nach Nahrung rufenden Institutes gedeckt werden.

Es ist nun freilich eine reine Vertrauenssache, dass man mir solche Mittel in die Haende gibt. Aber nachdem sogar der mit Recht misstrauischste und ex officio vorsichtigste Factor, der Staat, mit 20,000 [Thaler] sein Vertrauen bewiesen hat, nachdem das erste Arbeitsjahr bewiesen hat, dass ein solches Vertrauen gerechtfertigt war, ist, glaube ich, kaum mehr zu erwarten, dass berechtigtes Misstrauen bestehen sollte.

Wäre es Ihnen wohl möglich, Excellenz, durch Ihr Wort und Aufforderung in den Kreisen des Adels und der reichen Kaufleute der Ostsee-Provinzen zu einer solchen Subscription einen namhaften Beitrag zu verschaffen?

Das ist das eine grosse Anliegen.

Nun das Zweite. Die Zahl der vermietheten Tische muss durchaus erhöht werden. Ich habe 6 neue Tische eingerichtet und muss sie vermiethen, um die ganzen Betriebskosten aus den Erträgen der Miethssummen [*sic*] bestreiten zu können. Das Aquarium bringt 5000 [Thaler]; diese sind mir unentbehrlich zur Abzahlung der letzten 15,000 [Thaler] Schulden, und zur Anlage eines Reservefonds von mindestens 5000 [Thaler].

Vierundzwanzig vermiethete Arbeitstische bringen 12,000 [Thaler]; die regelmässigen Betriebskosten betragen 13,000 [Thaler]; also bei *grosser* Sparsamkeit kann die Aquariums-Einnahme einmal ausfallen, ohne dass der Bestand des Institutes sogleich in Frage gestellt wird. Die 24 Tische sind aber unerlässlich, und zwar schon im nächsten Jahr, d. h. vom October 1876 an. Diess habe ich sowohl in Rom wie in Berlin klar gemacht, und an beiden Orten volles und geneigtes Gehör gefunden. Der italienische Minister hat das Gutachten aller Universitäten Italiens eingefordert, und auf ihr einstimmiges Urtheil hin in sein nächstes Budget statt 2, 4 Tische gesetzt, und in Berlin ist mir versprochen, dass im nächsten Jahr ebenfalls eine Vermehrung der Tische stattfinden würde. Andrerseits sind meine Beziehungen in St. Petersburg sehr günstige; nicht nur mit Hrn. v. Bradke bin ich in freundlichem Zusammenhange und Correspondenz, sondern—was in diesem Falle wohl noch wichtiger ist,—mit dem Finanzminister selber, also mit Herrn von Reutern habe ich mancherlei Briefe gewechselt, und ich weiss, dass Se. Excellenz an der Zool. Station ein persönliches Interesse nimmt, und bereits darauf vorbereitet ist, dass [eine?] Mehrforderung an ihn gelangen wird.

Nach Ueberlegung erscheint es mir, dass Sie, Excellenz, die Verdoppelung der Tische, falls Sie zu solchem Schritte überhaupt geneigt wären, durch Ihren rein persönlichen Antrag bei beiden Ministerien durchsetzen würden. Ich selber kann den Antrag nicht stellen. Die vielen, sich widerstrebenden Russischen Gelehrten, die Gesellschaften u. Universitäten unter einen Hut zu bringen, würde wieder eine herzlich schwierige, höchst mühsame und vielleicht doch schliesslich vergebliche Anstrengung sein,—und ob eine solche Gesammt-Erklärung, falls sie zu Stande käme, den Nachdruck hätte, den,—bei den in Rede stehenden Persönlichkeiten—Ihr einzelnes Votum hat, ist wohl sehr zweifelhaft. Es kommt dazu, dass von den russischen Gelehrten und Universitäten formelle Proteste gegen die stipulirte Schliessung der Zool. Stations-Laboratorien vom 20 Juni–20 August unsres Kalenders erfolgt sind. Ich wäre bereit, für Russland diese Ferien abzuschaffen, falls mir für meine Concession mit einer Vermehrung der Tische von 2 auf 4 geantwortet würde,—hiedurch also würde das Ministerium auch den Wünschen der eignen Nationalen entgegenkommen.

Ich trage Ihnen Alles diess so brevi manu vor, Excellenz, [weil?] ich dreist genug bin zu glauben, ich bedarf bei Ihnen keiner weiteren Emballage meiner Anliegen; Sie werden ja Selbst am besten wissen, ob, was ich erbitte, im Bereiche Ihrer Macht und auch in Ihrem Willen liegt.

Zögern darf ich nicht. Das Schicksal der Agassiz'schen Zoologischen Station zeigt deutlich, dass periculum in mora ist,—andrerseits bin ich zu fest von der Lebensfähigkeit der Station überzeugt, und setze zu grosse Hoffnung auf Ihre wissenschaftliche Leistungsfähigkeit, als dass ich nicht alle Kräfte aufböte zur Vervollständigung ihrer Ausrüstung.

Oft auch beflügelt die Erwägung meine Maassnahmen, dass vorderhand das junge Institut meiner persönlichen Kräfte nicht entbehren kann, trotzdem ich alle Maassregeln so treffe, die Zool. Station fähig zu machen, auch meinen plötzlichen, durch Typhus, Cholera oder sonstigen Unglücksfall herbeigeführten Tod überleben zu können. Wenn es *mir* grosse Schwierigkeiten macht, die Sache bis ans Ende durchzuführen, so würde es jedem Andern fast unmöglich werden, es sei denn er würfe von Neuem bedeutende eigne Mittel in die Wagschale des Gelingens. Und bis dahin sehe ich Niemand, der dazu Neigung hätte,—wohl aber Viele Kleingläubige, Zweifler, Bekrittler und wie man die minores gentes bezeichnen mag.

Es ist mir sehr bezeichnend, dass ich unter den Angehörigen vergangner Zeiten und vergangener Geisteströmungen eigentlich mehr hingebende und thatkräftige Unterstützung gefunden habe, als unter meinen Altersgenossen. Vielleicht erklärt sich das aber mit Goethe's beruhigenden Worten:

> Lebst im Volke; sei gewohnt
> Keiner je des Andern schont.

Und da sieht man denn oft, dass man selbst der Andern nicht schont, und wundert sich nicht, wenn man allerhand Widerstaenden begegnet.—

Bis Ende September gedenke ich noch in Deutschland zu bleiben und soweit es in meinen Kraeften steht, persönlich zu einem guten Erfolge der Subscription beizutragen. Die Station ist in sehr guten Haenden und an sich, wenn einmal alle pecuniären und organisatorischen Schwierigkeiten gelöst sind, bietet die Verwaltung des Institutes gar keine sehr bedeutenden Schwierigkeiten.—

In der Hoffnung, Excellenz, dieser Brief werde Sie in völlig ungetrübtem körperlichen Wohlsein wieder antreffen, und eine freundliche Aufnahme finden, grüsse ich mit all der herzlichen Ergebenheit, die ich immer empfunden und mit der schuldigen Ehrfurcht zu verbinden gesucht habe, die der junge Adept dem unerreichbaren Meister nicht nur schuldet, sondern aus innerstem Bedürfniss darbringt.

Sollte der Sommer oder Herbst mich nicht mehr bis in die Breitengrade Dorpat's führen, so rechne ich doch mit ziemlicher Sicherheit auf nächstes Frühjahr, um eine Reise durch Russland viâ Constantinopel Odessa anzutreten und bis Petersburg u. Dorpat auszudehnen.

Mein Vater grüsst wie [*sic*] jedenfalls mit der herzlichsten Theilnahme und hofft auf gute Nachrichten von Ihnen.

In derselben Hoffnung schliesst diesen Brief
Ew. Excellenz
treu anhaenglicher
Anton Dohrn

---

[35] von Baer an Dohrn
16./28.8.1875, Dorpat

Dorpat d.16 August
resp. 28 Aug. 75.

Hochgeehrter Herr Doctor!

Obgleich ich nicht weiß, wie u[.] wo ich Sie treffen kann, beeile ich mich doch, Ihren lieben u[.] geehrten Brief recht bald zu beantworten. Ich will diese Antwort nach Stettin adressiren in der Hoffnung, daß sie von da aus Ihnen zugestellt wird.

Leider ist das, was ich zu sagen habe sehr wenig tröstlich. Eine irgend bedeutende Subscription hier im Lande halte ich für ganz unthunlich. Vom Adel ist die bei weitem größere Hälfte verarmt. Es gibt zwar einige Reiche u[.] auch sehr Reiche darunter, allein der Gesammt-Adel hat jetzt die [*sic*] sehr laut werdenden Anforderungen zur Errichtung von Volksschulen zu genügen u[.] bringt in dieser Sphäre namhafte Opfer. Eine Korporation hat allerdings in der letzten Zeit bedeutende Geldmittel gewonnen, nemlich [*sic*] die Revalsche Kaufmannschaft, welcher die Baltische Eisenbahn einen großen Geschäftskreis eröffnet hat. Allein, daß die Revalsche Kaufmannschaft irgend jemals für eine wissenschaftliche Anstalt [sich] interessirt hätte, habe ich nie gehört. Hier in Dorpat ist eine Gesellschaft für naturhistorische Erkenntniß der Baltischen Provinzen, von der ich die Ehre habe der sehr unnütze Präsident zu sein. Diese Gesellschaft laborirt seit Jahren am Verscheiden aus Mangel an Mitteln. Noch niemals aber hat ein Kaufmann aus Reval oder aus dem reichen Riga einen Beitrag gegeben. Sie fristet das Leben durch Beiträge von Professoren und Schullehrern.

Aber auch der andere Weg, die Subscription für russische Naturforscher zu verdoppeln, ist mir versperrt. Ich hatte den Minister des Unterrichts um eine Subvention für die hiesige naturforschende Gesellshaft gebeten u[.] erhielt eine abschlägige Antwort, obgleich allen übrigen russischen Universitäten für die Gründung naturhistorischer Gesellschaften diese Subvention gewährt war. Ihre eigenen Schritte können viel eher von Erfolg sein. Wenn Sie Rußland gehörig kennten, würden Sie wissen, daß man einem Ausländer viel eher eine Bitte derart gewährt, als einem Inländer. Man will im Auslande gelobt sein. Darauf gründen sich die Sinaitischen Bibeln u[.] ähnliche Ereignisse. Ueberdies muß ich mittheilen, daß der Doctor Rosenberg hierher geschrieben hat, er habe den bestellten Tisch nicht vorbereitet gefunden, sondern erst später erhalten. Er hat das nicht mir geschrieben, obgleich durch meine Korrespondenz mit Hf. v. Bradke ihm die Bewilligung zugekommen war. Eben deswegen hat man mir die Klage gleich hinterbracht. Ob sie nicht auch zu Bradke gedrungen ist, weiß ich nicht. Könnten Sie nicht, um Ihr Gesuch in Petersburg zu unterstützen, sich an einen derjenigen Russen wenden, welche Ihre Anstalt besucht u[.] benutzt haben. Ein solcher würde am nachdrücklichsten die Vortheile schildern können, die ein Fremder bei Ihnen vor jedem isolirt ankommenden Naturforscher hat.

Mit dem herzlichen Wunsche, daß Sie in diesem "non possumus" nicht Gleichgültigkeit, sondern nur einige Einsicht in die Verhältnisse vermuten mögen u[.] mit der Bitte, mich ferner in freundlichem Andenken zu bewahren,

Ihr
ergebenster
Dr K E.v Baer

---

[36] Dohrn/CAD an von Baer
3./4.9.1875, Stettin

Stettin. 3. September 1875.

Ihr freundlicher Brief, Excellenz, erwartete mich hier, als ich gestern von einer kurzen Reise nach Kopenhagen und Stockholm wieder bei den Meinigen eintraf. Ich war nach Scandinavien gegangen, um wie überall auch dort für mein Unternehmen zu sorgen, und hoffe, dass es den dortigen Naturforschern gelingen wird, mir Beistand zu gewinnen.

Was Sie mir über die Unmöglichkeit schreiben, in Ihrem Lande etwas zu machen, begreife ich nur allzu gut. Auch entmuthigt mich das nicht, ich bin es gewohnt, auf solche Schwierigkeiten besonders in dieser finanziell schwierigen Zeit zu stossen. Nur unverzagte Beharrlichkeit kann mir meinen definitiven Sieg gewinnen, und ich brauche nicht zu betheuern, dass ich sehr entschlossen bin, diesen Sieg à tout prix mir zu sichern.

Dass kleine Aergerlichkeiten auf Schritt und Tritt sich einstellen, kann ich auch nicht hindern, und so kann ich auch z. B. Herrn Dr. Rosenberg nicht verbieten, Dinge zu sagen und zu schreiben, die er loyaler und wahrheitsgetreuer entweder ganz unterdrückt oder mit allen verursachenden Momenten dargestellt hätte. Genüge es aber, Ihnen zu sagen, Excellenz, dass Herr von Bradke weit entfernt ist, mir die Schuld dieser Unannehmlichkeit zuzuschieben, da er sehr genau weiss, dass es die Schuld des Ministeriums war. Folgendes schreibt er mir in einem sehr liebenswürdigen, fast freundschaftlichen Briefe vom 2/14 Juli darüber: "– – ich hoffe, dass aehnliche Zeit-Conflicte in der Tisch-Benützung nicht mehr vorkommen werden. Die, welche stattgefunden, lassen sich auf einen Personen-Wechsel in meinem Departement zurückführen, der von mir vorgenommen werden musste.–" Das war die Antwort auf eine Bemerkung meinerseits. Wollte Herr Dr. Rosenberg davon Notiz nehmen, so würde er nur thun was ehrenhaft ist.–

Ich bin durchaus davon überzeugt, dass es vielleicht am klügsten ist, wenn ich wieder persönlich in St. Petersburg meine Sache führe, und denke Ende October nach dorthin zu reisen. Dann werde ich auch die Freude haben dürfen, Sie in Dorpat zu besuchen, – und das ist ein so alter Wunsch, dass ich seine Erfüllung nur hinter die dringenden Pflichten schob, die mich oft zwingen, Alles Andre hintenanzusetzen, um erst und vor Allem die finanziellen Schwierigkeiten der Zool. Station zu überwinden. Sie werden dann vielleicht so freundlich sein, Excellenz, mir noch einige gute Winke zu geben, und ich kann Ihnen die Situation der ganzen Angelegenheit besser expliciren, als durch 10 Briefe. Ich freute mich, aus dem Schweigen in Ihrem letzten Briefe entnehmen zu dürfen, dass die Wunde am Bein, über die Ihr vorletzter Brief klagte, wohl geschlossen ist, und dass Sie jetzt wieder Ihre alte stolze Gesundheit haben. Möge Sie Ihnen recht, recht lange noch bleiben!

Mit diesem herzlichsten Wunsche empfiehlt sich

Ihr
treu ergebener
Anton Dohrn

WR [?] Ich bleibe fortdauernd in Deutschland bis ich Alles durchgeführt habe. Die Station ist in guten Haenden.

[CAD:]
Excellens Ursa maxima
die ungeziefer besessne Welt wird nichts dabei verlieren, wenn ich einen gelehrten Artikel über australische Paussiden und darin ein sauersüßes lamento über Freund Westwood's Bockschüsse in seiner Monographie unterbreche, um Ihnen hinter das illustre Gaudissart de la Station Zodiacale noch einen herzlichen Gruß und Händedruck zu autographiren. Gestern hatte ich Anfälle von ächt englischer high aristocracy, nehmlich ein leiches Podagra, heute ists vorüber, dafür aber morgen meiner Frau Geburtstag, zu welchem ich ihr einen Pelz schenken will, ohne ihr die dazu gehörige Laus hinein zu setzen – es ist nichts vollkommen auf dieser lausigen Erde als die Ergebenheit, mit welcher Sie grüßt

Ihr allzeit unveränderter
C. A. Dohrn

4 Septb 1875.

[37] von Baer an Dohrn
17.9.1875, Dorpat

Dorpat d. 17. Sept. 1875.

An Herrn Dr. Dohrn jun.

Hochgeehrter Herr Doktor!

Ungeachtet der Hoffnung, Sie hier begrüßen zu können, was mir ein Lichtpunkt in meiner Einsamkeit sein würde, fühle ich das Bedürfniß, Ihnen vorher zu schreiben, um einen unbegründeten Verdacht, den ich leicht erregt haben könnte, zu heben. Ich will gleich zur Sache übergehen, da die Einleitungen noch mehr Wichtigkeit und Dunkelheit in die Sache bringen würden. Ich hatte geschrieben, daß Dr. Rosenberg hierher angezeigt habe, daß er keinen Arbeitstisch vorgefunden habe. Damals war Dr. Rosenberg noch nicht zurück. Wenige Tage später kam er an, und theilte mit, daß durch Schuld des Ministeriums Sie von der Verfügung desselben keine Nachricht erhalten hatten, wie Sie selbst in Ihrem letzten geehrten Brief mir mittheilen.

Auch hatte Dr. Rosenberg nicht gegen mich sich darüber beschwert, daß er nicht Alles zu seinem Empfang vorbereitet gefunden habe. Seine hierher gemachte Mittheilung kam mir nur von der Seite zu Ohren und ich hatte allerdings vorausgesetzt, daß er sich beschwert hatte. Seine Beschwerde konnte aber nur das russische Ministerium treffen. Sie haben also seine Bedenken nicht carbone in Ihrem Herzen zu notiren.

Mein Geschwür am Fuße ist noch immer nicht vollständig geheilt, aber der Heilung so nahe, daß nur noch eine [*sic*] Rest von der Größe eines Stecknadelkopfes übrig sein soll. Ich selbst kann leider nicht so weit sehen.

Ihren Bericht über den Fortgang der Stazione in der Zeitschrift für wissenschaftliche Zoologie habe ich im Separatabdruck dankbar empfangen.

Mit herzlichem Gruß bis auf ein künftiges Zusammentreffen

Ihr ergebenster
Dr K E v Baer

[38] Dohrn an von Baer
15.10.1875, Berlin

Berlin. 15. October 1875.

Excellenz!

Es sieht bös aus mit meinem Besuch bei Ihnen. Ich arbeite wie ein Pferd, – nein wie eine Dampfmaschine an der Hervorbringung von Mitteln für die Zool. Station, hoffe auch auf Erfolg, – aber nur bei rücksichtslosester Ausdauer und Beharrlichkeit. So verschiebt sich meine Abreise von Tag zu Tag. Moskau habe ich schon aufgegeben, da ich am 6ten November in Berlin einen Vortrag in der geographischen Gesellschaft halten muss, der die Subscription durch ganz Deutschland fördern soll. Aber nach Petersburg gehe ich à tout prix, um 2 weitere Tische zu vermiethen. Damit wäre die Sache auf lange abgeschlossen und so viel Mittel zu meiner Disposition um das ganze Institut, das in der deutschen Regierungs- wie wissenschaftlichen Welt sich einen vollen Erfolg errungen hat, durch und durch zu entwickeln. Ich denke das hängt wesentlich vom Finanzminister v. Reutern, und von Herrn v. Bradke ab. Mit beiden Herren habe ich persönliche, vertrauliche Beziehungen, so dass ich sofort in Petersburg drauf los gehen kann. Aber, Excellenz, seien Sie nicht böse, wenn ich trotz Ihres letzten Briefes doch darauf zurückkomme, dass Ihre Intervention mir die Wege noch weiter ebnen kann. Ich *weiss*, dass bei beiden genannten Herren Ihr persönlicher Einfluss grösser ist, als der der Akademie, und sämmtlicher National-russischen Forscher. Falls es Ihnen also nicht zu beschwerlich ist, sowohl Hrn. v.Bradke wie Herrn v.Reutern zu schreiben, so bin ich überzeugt, dass Sie mit der einfachen Bitte, wenn es irgend thunlich sei, meine demnächst persönlich vorgebrachten Anträge anzunehmen, mir sicherlich den Erfolg erleichtern würden. Ich würde mich natürlich verbindlich machen, davon überall zu schweigen,—was ich ja recht oft thun muss, um mir meinen Einfluss zu erhalten.

Dann würde auch vielleicht mein Aufenthalt in Petersburg sich so kürzen lassen, dass ich auf dem Rückwege in Dorpat einpassiren könnte, wonach ich so lange und so aufrichtig Verlangen trage.

Die Anmeldungen für diesen Winter in der Zool. Station sind wieder zahlreich, unter Andern Carpenter, Pringsheim[,] Victor Hensen aus Kiel ecc. Auch die Schweiz wird wahrscheinlich jetzt beitreten.

Meine Addresse in St. Petersburg wird sein bei "Alexander v. Baranowski. Fantanka dom Fedoroff", dem Onkel meiner Frau.

Die Aufnahme meiner Schrift über den Functionswechsel ist eine bei weitem günstigere als ich erwartete, hier in Berlin sowohl wie in Leipzig und vor Allem auch in England. Dort hat man sogar die *Absicht*, sie zu übersetzen,—ob es dazu kommt, weiss ich freilich nicht. Ich wünsche sehnlichst, die practische Arbeit bei Seite legen zu können, um die theoretische zu ergreifen, und die Hypothesen, die ich dort ausgesprochen habe, zu beweisen. Hoffentlich gelingt das mit Ausgang dieses Winters. Ein Paar Jahre brauche ich freilich, aber dann sollen Dinge ans Tageslicht kommen, die manche offne Mäuler nach sich ziehen werden. Q. d. b. v.!

Leben Sie wohl, Excellenz, hoffentlich sehe ich Sie doch noch und darf Ihnen dankbar die Hand drücken.

Ihr treu ergebner
Anton Dohrn

[39] Dohrn an von Baer
15.10.1875, Berlin

Berlin. 15ten October 1875

Excellenz!

In meinem heutigen Brief habe ich vergessen Ihnen mitzutheilen, dass sowohl die preussische Regierung wie die Berliner Akademie zur Herausgabe der hinterlassenen Schriften von Casp. Friedr. Wolff mir Mittel bewilligen wollen,–einstweilen freilich sub rosa!

Ich werde also den Grafen Lütke aufsuchen und ihm mit Ihrer Genehmigung meinen Plan vorlegen, um ein provisorisches Abkommen zu ermöglichen. Ich würde dann eine Gesammt-Ausgabe von Wolff's Schriften veranstalten, da Vieles ganz verschwunden ist. Darüber würde ich mich freuen, Ihren Rath und Meinung zu kennen.

Herzlich grüssend und in anhaenglichster Ergebenheit

Ihr
Anton Dohrn

[40] Dohrn an von Baer
22.10.1875, St. Petersburg

Fantanka 13. Dom Fedoroff.
St. Petersburg. 22. Oct. 75.

Excellenz!

Von Tag zu Tage verschob ich meine Abreise von Berlin, bis zum 19ten Abends, wo ich dann direct bis hierher ging. Wie ich hier fertig werden kann, weiss ich vorderhand nicht, denn am 4ten November muss ich wieder in Berlin eintreffen, um am 6ten einen Vortrag in der Geographischen Gesellschaft zu halten.

Sie werden mich hoffentlich nicht zu sehr schelten, dass ich rücksichtslos grade auf mein Ziel losgehe,–kostet es mich doch selbst oft die schwerste Resignation. Sie zu sehen und mit Ihnen sprechen zu können, ist ein so alter Wunsch, dass ich förmlich anfange zu glauben, das Schicksal wolle mir mit so vielem Gelingen auch ein empfindliches Misslingen bescheeren, da es mir zum zweiten Male die Aussicht dazu versperrt.

Und dass ich so lange in Berlin aufgehalten wurde, hatte seinen Grund in der Schwierigkeit des tot capita tot sensus. Da ich wieder mit der Akademie zu verhandeln hatte, musste ich die Mitglieder der math-physik. Classe der Reihe nach, mehrere sogar mehr als einmal besuchen, ehe ich mir sagen durfte, dass ich soviel gethan hätte, um einigermaassen auf Erfolg rechnen zu koennen. Dann war noch das Unterrichts-Ministerium da, und obschon es dort immer viel rascher geht, und die Dispositionen mir durchaus wohlwollend und hilfreich sind, so verzögerte sich mancherlei durch Abwesenheit einer wichtigen Persönlichkeit. Es handelt sich für die Station um nichts geringeres als den lang begehrten kleinen Dampfer, und es scheint begründete Hoffnung zu bestehen, dass in Berlin das Geld dazu hergegeben wird.

Wie es mir nun hier in Petersburg gehen wird, weiss ich nicht. Hr. v. Bradke, den ich heut um eine Unterhaltung bitten liess, ist nicht en ville, kommt erst Sonntag. Er wird mir sagen, wie die Dinge liegen. Bei Ihrem Herrn Collegen Brandt erfuhr ich, dass in der That die Akademie gar nicht weiter über die Sache informirt ist, wie denn Prof. Brandt nichts davon wusste, ob überhaupt russische Zoologen in der Station gearbeitet haben. Hieraus ist also wohl auch zu schliessen, dass die weitere Behandlung der Angelegenheit ausschliesslich in den Haenden des Ministeriums liegt, und dass in letzter Instanz wohl Hr. v. Reutern das letzte Wort redet. Das wird sich ja wohl bald herausstellen, und ich werde nicht verfehlen, Ihnen, Excellenz, davon Mittheilung zu machen. Bis dahin nehme ich Urlaub, und füge nur noch die Bitte hinzu, dass Sie die grosse Güte hätten, etwa bei Ihnen für mich eingetroffene Briefe freundlichst an meine hiesige Addresse senden zu wollen.

Hoffentlich höre ich zugleich mit zwei Worten die Bestätigung Ihrer fortdauernden Gesundheit, und da man ja nie aufhören soll zu hoffen, so hoffe ich auch, dass was 1874 u. 75 mir versagt haben, 1876 erfüllen wird, und dass ich dann endlich sehen kann, ob die Büste, die von Ihnen im Saal der Zool. Station errichtet ist, dort auch wirklich Sie darstellt.

Mit herzlichem Grusse
Ihr
treu ergebner
Anton Dohrn

---

[41] Dohrn an von Baer
2.5.1876, Neapel

Napoli. Stazione Zoologica.
2. Mai 1876.

Excellenz!

Die aeusserste Geschäftigkeit, die mich wiederum in diesem Frühjahr überfallen, hat mich bisher abgehalten, Ihnen für die Uebersendung Ihres Buches zu danken.

Ich brauche nicht zu wiederholen, mit welchen Empfindungen ich Ihre Darlegungen gelesen habe. Das historische Auffassen, das so mächtig aus Ihren Worten hervordringt, wird ja freilich immer geringer;

man vergisst über der Praehistorie jetzt nur allzu leicht die Historie, und dass man selbst mit Denken und Wollen ein geschichtliches Atom ist, dessen Gegenwart ebenso wenig bleibend ist, wie irgend etwas Andres. Kommt dann plötzlich eine lebendige Stimme und spricht als Augenzeuge von vergangenen Epochen, die wie Tertiärgesteine von uns entfernt sind, und nur durch einige versteinerte Begriffe uns einen Beweis von einstmals pulsirendem geistigen Leben geben, so fasst uns wohl das Bewusstsein, dass wir uns zur Zukunft verhalten, wie jene zu uns, – und wir lernen besser begreifen, dass auch wir in's Bodenlose versinken werden mit all den geträumten Besitzthümern von Einsichten u. Wahrheiten, die erst den Beweis zu liefern haben, ob sie eine geistige Epoche überdauern können.

Ich habe Ihnen noch persönlich zu danken, dass Sie die Vorstellungen, die in meinem Kopfe gekeimt haben, in Ihrer Nachschrift einer so nachsichtigen Beurtheilung unterworfen haben. Prof. Semper hat das zwar übelgenommen, aber diese Prioritäts-Angst ist wohl auch ein Zeichen unhistorischer Bildung, wenn nicht schlimmerer Bildungslosigkeit. Es erscheint gradezu unfassbar, dass heutzutage Ueberzeugungen Jahre lang bestehen sollten, ohne durch ein halbes Dutzend Vorläufiger und ebenso viel Nachläufiger Mittheilungen seicht und schal gemacht zu werden. Wie verschieden von den Zeiten, da Ihr Verleger Sie vor dem Publicum verklagte, er könne Sie nicht bewegen, den 2ten Band der Entwicklungsgeschichte abzuschliessen! Die Vorläufigen Mittheilungen sind wie die neuen Gewehre, die in einer Minute sieben Mal abgeschossen werden koennen, – man zielt darum auch weniger sorgfältig. –

Von der Zoologischen Station kann ich nur Gutes melden. Nachdem sie anfänglich mit vielleicht übertriebenen Erwartungen begrüsst war, [erfolgte?] eine Reaction, und zumal in Deutschland wucherten allerhand böse Klä[tsche]reien, die meist aus persönlichem Uebelwollen entsprangen. Ich konnte dagegen nichts machen, als unbeirrt fortfahren, das Institut auszubauen. Dieser Winter hat nun Forscher von reiferem Urtheil hierhergeführt, und ich habe die Freude gehabt, dass Proff. Hensen aus Kiel, His aus Leipzig, Grenacher aus Rostock, der alte Carpenter aus London, ferner der Botaniker Reinke aus Göttingen und Mehrere Andere mir versichert haben, sie seien Alle höchlichst befriedigt und würden privatim bei den Regierungen u. öffentlich in den Zeitschriften ihr günstiges Urtheil über die Station dem schleichenden Geklätsch entgegenstellen [?]. Damit wäre es denn ein für alle Mal mit den Versuchen, mir ein Bein zu stellen, vorbei.

Ich habe jetzt den Ersten JahresBericht veröffentlicht, der hauptsächlich Rechenschaft ablegt über die Verwaltung des Instituts. Demnächst wird dasselbe wesentlich bereichert werden durch den Besitz eines kleinen Dampfschiffs, für dessen Ankauf Seitens der Berliner Akademie und des Unterrichtsministers 8000 Thaler bewilligt worden sind. Auch hat die innere Organisation der Anstalt Fortschritte gemacht, das Personal ist geschult, Routine tritt an Stelle persönlicher Initiative, und es wird allmälig möglich, die rein geschäftliche Thätigkeit unter die wissenschaftliche herabzudrücken. Viele wichtige u. einflussreiche Menschen haben das ganze Institut in diesem Frühjahr besichtigt, es wächst an Popularität, so dass ich hoffe, es werde mir jetzt leichter gelingen, die noch nöthigen Mittel zu erwerben.

Leider ist dies Jahr kein Russe hier gewesen. Dadurch fällt für mich die Möglichkeit die Russische Regierung um Vermehrung der Tische zu bitten, was mir schon vom Finanz Minister Hrn. v. Reutern zugesagt war. Es ist das sehr schade, denn das Institut braucht ein Einkommen von 15–16000 [Thaler] und muss diese Summe zumeist aus den Tisch-Miethen zu erwerben trachten. Woran es liegt, dass plötzlich die Russen ausbleiben, weiss ich nicht.

Ich erlaube mir mit diesem Briefe einige Plaene der Station zu übersenden. Wenn Sie auch, Excellenz, dieselben nicht weiter ansehen und daraus die Einrichtung des Institutes erkennen könnten, so hoffe ich doch, dass Ihnen von Angehörigen ein Bild der Sache nach der Lectüre des JahresBerichts gegeben werden kann, und es wird mich wie immer mit Stolz erfüllen, wenn ich weiss, dass Sie Ihren Antheil und Ihre Sympathie diesem Unternehmen fortgesetzt bewahren. Der nächste Jahres-Bericht soll dann schon mehr wissenschaftliche Früchte bringen, die das Institut in sich selbst zeitigt.

Mit unveraenderlicher Verehrung
Ihr
Anton Dohrn

BIBLIOGRAPHY

Agassiz, E. C., ed., *Louis Agassiz. His Life and Correspondence.* Boston, New York: Houghton, Mifflin and Co., 1900.

Baer, K. E. von, *Ueber Entwickelungsgeschichte der Thiere. Beobachtung und Reflexion.* Part one: Königsberg: Gebr. Bornträger, 1828, XXII, 271 pp., 3 copper plates. Part two: Königsberg: Gebr. Bornträger, 1837, 315 pp., 4 copper plates.

———, *Reden gehalten in wissenschaftlichen Versammlungen und kleinere Aufsätze vermischten Inhalts.* St. Petersburg: H. Schmitzdorff. Vol. 1: *Reden.* 1864; vol. 2: *Studien aus dem Gebiete der Naturwissenschaften.* 1876; vol. 3: *Historische Fragen mit Hülfe der Naturwissenschaften beantwortet.* 1873.

(———), *Das fünfzigjährige Doctor-Jubiläum des Geheimraths Karl Ernst von Baer am 29. August 1864.* St. Petersburg: Kaiserliche Akademie der Wissenschaften, 1865, 128 pp.

———, *Nachrichten über Leben und Schriften des Herrn Geheimraths Dr. Karl Ernst von Baer, mitgetheilt von ihm selbst.* St. Petersburg: Ritterschaft Ehstlands, 1864, distributed privately; trade editions: St. Petersburg: H. Schmitzdorff, 1866; Braunschweig: F. Vieweg, 1886. English translation of the 1886 edition: K. E. von Baer, *Autobiography of Dr. Karl Ernst von Baer,* ed. Jane M. Oppenheimer, trans. H. Schneider. Canton, Mass.: Science History Publications, 1986, XIV + 389 pp.

———, "Die Zoologische Station in Neapel." *St. Petersburger Zeitung,* Nr. 219 (20.8./1.9.1873).

———, *Ueber Darwin's Lehre.* In K. E. von Baer, *Reden* vol. 2: *Studien aus dem Gebiete der Naturwissenschaften.* St. Petersburg: H. Schmitzdorff, 1876, pp. 235–480.

Dohrn, A., *Kurzer Abriss der Geschichte, sowie Gutachten und Meinungsäusserungen hervorragender Naturforscher über die Gründung der Zoologischen Stationen.* Jena: Frommann, 1871, 8 pp.

———, *Der Gesammt-Organisationsplan der Zoologischen Stationen.* N. pl. (September 1871), 2 pp.

———, "Der gegenwärtige Stand der Zoologie und die Gründung zoologischer Stationen." *Preuss. Jb.* 30 (1872): 137-161. Reprinted in *Naturwiss.* 19 (1926): 412–424; and in H.-R. Simon, *Anton Dohrn und die Zoologische Station Neapel* pp. 23–46. Italian translation in *Nuova Antologia* Jan. (1873): 1–27. Reprint in: *Boll. zool.* 35 (1968): 507–531.

———, *Die Bibliothek der Zoologischen Station Neapel. Verzeichniss der daselbst bis zum Ende 1873 vorhandenen Bücher.* Leipzig: Engelmann, 1874, 91 pp.

———, *Der Ursprung der Wirbelthiere und das Prinzip des Functionswechsels.* Leipzig: Engelmann, 1875, 87 pp. English translation with an introduction by Michael T. Ghiselin, in preparation.

———, "Mittheilungen aus und über die zoologische Station von Neapel. Offenes Sendschreiben an Prof. Dr. C. Th. von Siebold." *Z. wiss. Zool.* 25 (1875): 457–480.

———, *Erster Jahresbericht der Zoologischen Station in Neapel.* Leipzig: Engelmann, 1876, 92 pp.

———, *Das 25-jährige Jubiläum der Zoologischen Station zu Neapel am 14. April 1897.* Leipzig: Breitkopf & Härtel, 1897, 44 pp. Reprint in Simon, *Anton Dohrn und die Zoologische Station*, pp. 59–104.

Dohrn, C. A., *Amtlicher Bericht über die 38. Versammlung Deutscher Naturforscher und Aerzte in Stettin.* Stettin: 1864.

Dohrn, H., "C. A. Dohrn." *Stettin. ent. Ztg.* 53 (1892): 281–322.

Dohrn, K., *Von Bürgern und Weltbürgern. Eine Familiengeschichte.* Pfullingen: Neske, 1983, 272 pp.

Fischer, J.-L., 1980. "L'aspect social et politique des relations épistolaires entre quelques savants français et la Station zoologique de Naples de 1878 à 1912." *Rev. Hist. Sci.* 33 (1980): 225–251.

Groeben, C., ed., *Charles Darwin–Anton Dohrn. Correspondence.* Naples: Macchiaroli, 1982, 118 pp.

———, ed., *Emil du Bois-Reymond–Anton Dohrn. Briefwechsel,* in collaboration with Klaus Hierholzer, with an introduction by Ernst Florey. Berlin, Heidelberg, New York, Tokyo: Springer-Verlag, 1985, 322 pp.

———, "Anton Dohrn–the Statesman of Darwinism." *Biol. Bull.* 168 (suppl. 1985): 4–25.

———, "The *Vettor Pisani* Circumnavigation (1882 to 1885)." *Dt. hydogr. Z.*, Ergänzungsheft B 22 (1990): 220–234.

Groeben, C. and Müller, I., *The Naples Zoological Station at the Time of Anton Dohrn.* Exhibition Catalogue. Naples: Stazione Zoologica, 1975, 110 pp.

Groeben, C. and Wenig, K., eds., *Anton Dohrn und Rudolf Virchow. Briefwechsel 1864–1902.* Berlin: Akademie Verlag, 1992, 132 pp.

Hensen, V., "Die Zoologische Station in Neapel." *Leopoldina* 12 (1876): 141–144; 153–156.

Heuss, T., *Anton Dohrn.* Tübingen: R. Wunderlich, 1962. (1st ed.: Zürich: Atlantis, 1940; 2nd enlarged ed.: Tübingen: 1948). Italian abbreviated translation by C. Gundolf: *L'"Acquario" di Napoli e il suo fondatore Anton Dohrn.* Rome: Ed. Casini, 1959. English translation by L. Dieckmann: *Anton Dohrn. A Life for Science.* Heidelberg, Berlin, New York: Springer-Verlag, 1991, 401 pp.

Knorre, H. von and Schierhorn, H., "Karl Ernst von Baer (1792–1876). Eine ikonographische Studie." *Acta Hist. Leopoldina* 9 (1975): 227–268.

Kofoid, Ch. A., "The Biological Stations of Europe." *US Bureau of Education Bull.* 4 (1910): Washington, 1910, 360 pp.

Kühn, A., "Anton Dohrn und die Zoologie seiner Zeit." *Pubbl. Staz. Zool. Napoli* (1950) suppl., 250 pp.

Lukina, T. A., ed., *Perepiska Karla Bera.* 4 vols. Leningrad, Nauka, Leningrad Section, 1970, 1975, 1976, 1978.

Lurie, E., *Louis Agassiz. A Life in Science.* Chicago: University of Chicago Press, 1960.

Maienschein, J., "Agassiz, Hyatt, Whitman, and the Birth of the Marine Biological Laboratory." *Biol. Bull.* 168 (suppl. 1985): 26–34.

Müller, I., *Die Geschichte der Zoologischen Station Neapel von der Gründung durch Anton Dohrn (1872) bis zum ersten Weltkrieg und ihre Bedeutung für die Entwicklung der modernen biologischen Wissenschaften.* Habilitationsschrift, Universität Düsseldorf, Math.-Naturwiss. Fakultät, 1976, 444 + 301* pp.

———, *Nikolai Nikolajewitsch Mikloucho-Maclay. Briefwechsel mit Anton Dohrn* (Beiträge zur Ethnomedizin, Ethnobotanik und Ethnozoologie, IV). Norderstedt: Verlag für Ethnologie, 1980, 127 pp.

Oppenheimer, Jane M., Some Historical Backgrounds for the Establishment of the Stazione Zoologica at Naples. In M. Sears and D. Merriman, eds., *Oceanography. The Past*. New York: Heidelberg, Berlin: Springer-Verlag, 1980, pp. 179–187.

———, Science and Nationality: The Case of Karl Ernst von Baer (1792–1876). *Proc. Am. Phil. Soc.* 134 (1990): 75–82.

Partsch, K.-J., *Die Zoologische Station Neapel. Modell internationaler Wissenschaftszusammenarbeit* (Studien zu Naturwissenschaft, Technik und Wirtschaft im Neunzehnten Jahrhundert, 11). Göttingen: Vandenhoeck & Ruprecht, 1980, 369 pp.

Raikov, B. E., *Karl Ernst von Baer 1792–1876. Sein Leben und sein Werk.* German translation by H. von Knorre (Acta historica Leopoldina, 5). Leipzig: Johann Ambrosius Barth, 1968. 516 pp., illus. (Russian 1st ed., 1961).

Simon, H.-R., ed., *Anton Dohrn und die Zoologische Station Neapel* (Bibliographia et Scientia, 1). Frankfurt a.M.: Ed. Erbrich, 1980, 164 pp.

Skalova', O., "An Analysis of Geographical Mobility of Scientists and their Communications as a Component of their Working Conditions with Regard to the Naples Zoological Station." *Pubbl. Staz. Zool. Napoli* 39 (1975), suppl. 2, 126 pp.

Stieda, L., *Karl Ernst von Baer. Eine biographische Skizze.* Braunschweig: Vieweg, 1877; 2nd ed., 1886.

Uschmann, G., *Geschichte der Zoologie und der Zoologischen Anstalten in Jena 1779–1919.* Jena: VEB G. Fischer, 1959, 249 pp.

INDEX OF NAMES

Agassiz, Alexander, 49n, 83n
Agassiz, Louis, 4, 5, 40, 49, 59, 83, 85, 102, 110, 118, 138, 141
Anderson, John, 49n, 83n
Audouin, Jean Victor, 5
Baer, Alexander von, 8n
Baer, August Emmerich von, 8n
Baer, Auguste von, née von Medem, 32n
Baer, Hermann von, 8n
Baer, Karl Friedrich Julius von, 8n
Baer, Karl Heinrich von, 8
Baer, Ludwig von, 8
Baer, Magnus Ludwig Conrad von, 8n
Baer, Marie von, 8n
Baird, Spencer F., 5
Balfour, Francis Maitland, 61n
Balfour, Gerald, 61n
Baranowska, Catharina von, née von Timler, 63n
Baranowska, Helene von, 63n
Baranowska, Marie von. *See* Dohrn, Marie
Baranowski, Alexander von, 91, 146
Baranowski, Georg Ivanovich von, 17, 61n, 63n
Beer, Otto, 61n
Borell, Peter, 56n
Boveri, Theodor, 2n
Bradke, Emanuel von, 66 ,82, 84, 87, 88, 90, 92, 124, 138, 140, 143, 146, 148
Brandt, Johann Friedrich, 92, 148
Brehm, Alfred, 4
Bronn, Heinrich Georg, 19
Calderon de la Barca, Pedro, 71, 128
Carpenter, William Benjamin, 90, 94, 146, 149
Carus, Julius Victor, 76, 132
Chierchia, Gaetano, 18
Claus, Carl, 79, 135
Czermak, Johann Nepomuk, 52, 61n, 113
Czermak-Lümel, Marie, 52n
Danilevskij, Nikolaj Jakovlevich, 33n
Darwin, Charles, 2, 10, 19, 21-23, 25, 33n, 39, 56n, 61, 70, 77, 78, 80, 81, 102, 120, 127, 134, 136
Döllinger, Ignaz, 40, 103
Dohrn, Anna. *See* Wendt, Anne
Dohrn, Boguslav, 63n
Dohrn, Carl August, 1, 6, 7-10, 13, 14, 25, 27, 28, 31, 32, 33n, 34, 35, 38-40, 42, 51, 52, 71, 82n, 88, 89, 97-100, 102-104, 113, 128
Dohrn, Catharina, 63n
Dohrn, Harald, 63n
Dohrn, Heinrich, 6
Dohrn, Heinrich (Jr.), 1, 7, 9, 10, 25, 27, 28, 31-35, 38, 40, 42, 52, 71, 83, 97-102, 104, 113, 128, 138
Dohrn, Marie, née von Baranowska, 3, 17, 63n
Dohrn, Reinhard, 63n
Dohrn, Wilhelm, 7
Dohrn, Wolfgang, 63n
du Bois-Reymond, Emil, 25, 38n, 39, 62n, 102
Eisig, Hugo, 45, 47, 58, 63, 107, 109, 117, 121
Elena Pavlovna, Grand Duchess of Russia, 8
Engelmann, Wilhelm, 22, 74-76, 130, 131, 132
Falk, Adalbert, 62, 121
Fiedler, Conrad, 61n
Flexner, Simon, 14
Friedrich Wilhelm IV, King of Prussia, 8
Furtwängler, Wilhelm, 7
Gegenbaur, Carl, 38n
Geoffroy St. Hilaire Étienne, 62, 121
Goethe, Johann Wolfgang von, 31, 37, 50n, 85, 101, 141
Graf, H., 36
Grenacher, Hugo, 94, 149
Grube, Eduard, 38n
Haeckel, Ernst Heinrich, 4, 10, 19-21, 33n, 35n, 38n, 45n, 46n, 70

Hagen, Hermann August, 40, 102
Harting, Pieter, 48n
Heidtmann, Captain, 32, 97
Heinrich Prince Reuss, 17, 65, 70, 127
Helmholtz, Hermann von, 38n
Hensen, Victor, 90, 94, 146, 149
Herter, Christian, 14
Heuss, Theodor, 2, 7, 26
Hildebrand, Adolf von, 56, 57, 116
His, Wilhelm, 94, 149
Hoff, Ph., 67, 68n
Humboldt, Alexander von, 7, 9
Huxley, Thomas Henry, 22, 39, 54, 61, 63, 80n, 102, 115, 120, 121
John, Th., 68, 71n, 72
Kaavere, Vello, 8n
Keyserling, Count Alexander von, 8
Kleinenberg, Nikolaus (Nikolai N.), 17, 46, 58, 60, 63, 66, 69, 80, 108, 117, 118, 121, 124, 126, 136
Knorre, Heinrich von, 1, 58n
Kölliker, Albert von, 62n
Konstantine, Grand Duke of Russia, 18
Kowalewski, Alexander von, 17, 62n
Krohn, August, 4
Krusenstern, Ivan Feodorovich von, 8n, 9
Lacaze-Duthiers, P.J. Henri de, 4, 5, 46, 108
Lankester, Edwin Ray, 52n
Leuckart, Rudolf, 38n, 39, 62n, 102
Lincoln, Abraham, 19
Lingen, Karl Magnus von, 8n
Lloyd, William Alford, 39n, 51, 113
Lobianco, Salvatore, 14
Lütke, Feodor (Friedrich Benjamin Count), 8n, 17, 18, 91, 147
Lukina, Tatiana Arkadevna, 16
Marées, Hans von, 53, 56n, 61n
Matteucci, Carlo, 5
Mendelssohn-Bartholdy, Felix, 7, 8
Metchnikoff, Elie de (Ilja I.), 17, 62n
Michael, Grand Duke of Russia, 8
Micl(o)ucho-Maclay, Nicolai N., 3, 35, 37n, 100
Milne-Edwards, Henri, 5
Miljutin (Milutine), Count Dimitri Alekseevitch, 65, 70, 123, 124, 127
Müller, Gustav, 48n
Müller, Irmgard, 4
Müller, Johannes, 4, 20, 62n
Nicholas I, Czar of Russia, 8
Otto, Adolf Wilhelm, 79, 135
Owen, Richard, 80n
Owsjannikow, Philipp Vasilevitch, 47, 109
Panceri, Paolo, 40n
Pander, Christian Heinrich, 17
Pringsheim, Nathanael, 90, 146
Profumo, Giacomo, 95n
Quatrefages de Bréau, Jean-Louis-Armand de, 4, 5
Raikov, Boris E., 1, 8n, 16n
Rajewsky, 64n
Reinke, Johannes, 94, 149
Renard, Karl, 17, 51, 113
Retzius, Anders, 4
Reuss, Prince. *See* Heinrich Prince Reuss
Reutern, Michael Christoforowitch von, 17, 65, 82, 84, 90, 92, 95, 124, 138, 140, 146, 150
Robertson, David, 3, 35n
Rosenberg, Emil, 66, 82, 83, 87–89, 124, 138, 143–145
Salensky, Wladimir W., 62n, 64n
Sears, Mary, 3n
Selenka, Emil, 48n
Semper, Carl, 93, 149
Siebold, Carl Theodor von, 62n, 79, 82, 135, 138
Spallanzani, Lazzaro, 4
Stieda, Alexander, 1
Stieda, Ludwig, 1, 16n
Strauss, David, 71, 127
Sutt, Toomas, 16n
Tischendorf, Konstantin von, 87n
Tolstoi (Tolstoy), Count Dimitri Andreevitch, 16, 17, 47n, 48, 50, 51, 56, 58, 65, 82, 110, 111, 112, 116, 117, 122, 123, 138
Trinchese, Salvatore, 40n
van Beneden, Édouard, 4
van Beneden, Pierre-Joseph, 4, 39, 102
Veselovsky, Konstantin Stepanovich, 17
Virchow, Rudolf, 25, 40n, 62n
Vogt, Carl, 5, 18, 38n
Wendt, Anna, 7, 51n
Wendt, Gustav, 51n
Westwood, John Obadiah, 88, 144
Wilhelm II, German Emperor and King of Prussia, 8
Wolff, Caspar Friedrich, 17, 44, 66, 69, 91, 107, 124, 126, 147

www.ingramcontent.com/pod-product-compliance
Lightning Source LLC
Chambersburg PA
CBHW081139300726
48982CB00006B/1010

* 9 7 8 0 8 7 1 6 9 8 3 3 9 *